Design and Construction
of Water Wells

Design and Construction of Water Wells

A Guide for Engineers

Prepared by the
National Water Well Association

Jay Lehr
Scott Hurlburt
Betsy Gallagher
John Voytek

VNR VAN NOSTRAND REINHOLD
_____ New York

Copyright © 1988 by Van Nostrand Reinhold
Library of Congress Catalog Card Number 87-16208
ISBN 0-442-26907-2

All rights reserved. No part of this work covered by the copyright hereon may be reproduced or used in any form or by any means—graphic, electronic, or mechanical, including photocopying, recording, taping, or information storage and retrieval systems—without written permission of the publisher.

Printed in the United States of America

Van Nostrand Reinhold
115 Fifth Avenue
New York, New York 10003

Van Nostrand Reinhold International Company Limited
11 New Fetter Lane
London EC4P 4EE, England

Van Nostrand Reinhold
480 La Trobe Street
Melbourne, Victoria 3000, Australia

Macmillan of Canada
Division of Canada Publishing Corporation
164 Commander Boulevard
Agincourt, Ontario M1S 3C7, Canada

16 15 14 13 12 11 10 9 8 7 6 5 4 3 2

Library of Congress Cataloging-in-Publication Data

Design and construction of water wells.

"Prepared by the National Water Well Association."
Bibliography: p.
Includes index.
I. Wells.   I. National Water Well Association.
II. Lehr, Jay H., 1936-
TD405.D47 1987   628.1'14   87-16208
ISBN 0-442-26907-2

Contents

Preface / vii

1. Ground Water and Geology / 1
2. Ground-Water Flow / 17
3. Drilling Technology / 25
4. Water-Well Design / 72
5. Intake Design / 84
6. Constructing Water Wells / 105
7. Testing for Yield and Quantity / 133
8. Water-Well Development / 148
9. Water-Well Maintenance / 161
10. Consolidated Rock-Well Design / 178
11. Ground-Water Regions of the United States / 193

Bibliography / 223

Index / 225

Preface

In the seventh and eighth decades of the twentieth century, government, industry, and the public began to recognize the vast value of man's ground-water supplies—a resource that was once considered to be the private domain of the geologist and water-well contractor. Today scientists and engineers from related disciplines such as chemistry, biology, mathematics, and civil engineering are moving rapidly into the field of ground-water development. A need for a basic preparatory text on the nature of water wells, their construction, development, and operation has been widely recognized. This book is intended to fill the void for the professionals new to ground water and wells, who will be working on projects involving well construction or maintenance. It is not intended to be a complex, sophisticated engineering bible on the subject, as such texts already exist. Rather, it is intended to introduce the nongeologic scientist and engineer to the basic facts needed to understand the more complicated material prepared by experts. This book is simply written with minimal jargon and orderly development of subject materials. We trust that readers will be grateful for the approach taken and will be able to use this knowledge in projects involving water wells.

The book begins with a general discussion of ground water and geology to acquaint the reader with the dynamics and occurrence of ground water on the earth. Fundamental physical properties pertaining to ground-water flow are introduced in chapter 2. We then move to the engineering-oriented portion of the text with a discussion of the advantages and drawbacks of the different methods of well drilling. A chapter on well design follows, including full details on the concepts and concerns of well design. Chapter 5 is dedicated to intake design, a critical but often misunderstood aspect of every water well. Water-well construction techniques are described clearly and concisely in chapter 6, followed by a thorough description of how a well is developed after construction to maximize its yield. Pumping tests are then considered as methods to use wells to analyze aquifer characteristics. Well maintenance and rehabilitation are given significant emphasis in chapter 9. A special section on rock-well construction has been added as the literature is largely devoid of basic information on this subject. We trust you find the book an enjoyable adventure into a widely misunderstood and underappreciated science that will ultimately free the world of its overdependence on surface water.

1

Ground Water and Geology

From where does the water in wells come? Why can water be found a few feet beneath the dry surface soil in some places but not at several hundred feet below the surface in others? These questions and others involve underground water, water that geologists call *ground water*. Ground water fills the pores and cracks of soil and rock beneath the land surface. It comes to the surface as springs, and it swells the volume of streams and rivers by seeping into them from their beds and from along their banks. Ground water also can be obtained by drilling water wells.

Water on the earth's surface is one part of an unending global sequence of water movement known as the *hydrologic cycle*. Water moves from the world's oceans to the land surface through evaporation and precipitation, making its way back to the oceans as streams and rivers. Water can be stored in all three of its natural physical states during its cycle on earth—liquid, water vapor, and ice. The majority—97 percent—of the world's water is held as liquid in the oceans. Polar ice sheets and glaciers account for about 2 percent. Underground water or ground water amounts to about 0.6 percent of the world's water. Surface water in the form of rivers and lakes amounts to only about 0.1 percent. The remaining fraction of all the water is held in the atmosphere as water vapor. The total amount of water on the earth's surface does not change; it is merely recycled through its three different natural states.

The energy to drive the hydrologic cycle is supplied by the sun. The sun's solar radiation warms surface water supplies, causing water to evaporate and enter the atmosphere as water vapor. Once in the atmosphere, the water vapor is transported by winds, which result from temperature differences in warmer and cooler parts of the earth. As the water vapor is transported by winds to cooler areas within the atmosphere, it condenses and forms clouds. Eventually, the condensate saturates the air, causing rain or snow to fall to the ground.

Ground water and ground-water flow are affected by the hydrologic cycle. When precipitation falls on the land, much of the water runs off the land surface, flowing into rivers, lakes, and eventually the oceans. Some rainwater evaporates directly from the land surface back into the atmosphere. And some of the precipitation soaks into the soil and enters the ground-water system.

Some of the water that infiltrates the soil may be absorbed by the roots of growing plants, carried up to their leaves, and returned to the atmosphere by a process called *transpiration,* the release of water vapor by plant foliage. The remainder of the ground water that infiltrates the land surface is stored in the pore spaces of soil and rock. From within the ground, some water returns to the surface by capillary action and evaporates, some is delivered to plant roots, and much reappears on the surface at lower elevations as springs or by seeping into streams.

Pore spaces in the soil and rock provide natural passageways that allow precipitation to infiltrate the land surface. These passageways are the result of several different mechanisms. The passages may be soil cracks caused by previous drying events, small tubules left by the decay of grass roots, larger openings formed by burrowing animals, or openings caused by the alternate freezing and thawing of ice crystals.

Even dense consolidated rock allows some water to infiltrate. These water passages can be large, as in faults or fractures that separate the rock mass. Intergranular openings in some types of dense consolidated rocks also provide passages for water penetration.

Once water enters the subsurface, it is either stored or transmitted, depending on how deep it infiltrates. Water occurs in two distinct zones in the subsurface. The uppermost zone, with openings that contain both air and water, is referred to as the *unsaturated zone.* Below the unsaturated zone is the *saturated zone,* in which all the pore spaces are filled with water. Under typical conditions found in a humid climate, the unsaturated zone contains the *soil water belt.* The soil water belt extends from the ground surface down through the major root zone. Its thickness varies with soil type and vegetation. A small volume of the precipitation that infiltrates the soil water belt is held there against the force of gravity by capillary forces. Water molecules are attracted to individual soil and rock particles and held in the form of a clinging film or droplet. The amount of water present in the soil water belt depends on the amount, duration, and timing of precipitation.

Water that has infiltrated the soil water belt but cannot be held there continues to travel down into the subsurface in response to gravity. As it moves downward, the infiltrating water passes through an intermediate zone known as the *vadose zone.* This intermediate zone, the lower part of the unsaturated zone, varies in thickness depending on the amount of water present.

The saturated zone lies directly below the vadose zone and marks the region where all of the pore spaces between the soil and rock particles are filled with water. The water found in the saturated zone is ground water and is the source of drinking water for more than 50 percent of the population of the United States.

The upper surface or top of the saturated zone is called the *water table.* Capillary tension draws water from the top of the saturated zone upward into the pore spaces of the overlying unsaturated zone to form the *capillary zone* or *capillary fringe*. This action partially fills the available pore space to a height ranging from less than an inch to as much as 2 feet or more depending on the geologic materials present. Usually finer-grained material will allow a thicker capillary fringe to develop.

Water is held above the water table in the capillary fringe only temporarily. Whenever heavy rains or massive snow melts provide large amounts of water for infiltration into the ground, the upper soil and rock layer, normally unsaturated, is temporarily saturated while the majority of the water moves down to the water table. Ground water is replenished through this process, referred to as *recharge*.

The water table elevation changes during recharge events. The water table will rise with respect to the land surface, depending on the amount of infiltrating water. Conversely, in periods of drought, the water table elevation decreases relative to the land surface. The relative position of the water table can be determined by noting the water level that stands in wells that penetrate the saturated zone. If many water wells have been drilled in a given area, a configuration of the local water table can be shown by drawing a contour map based on the levels of standing water in the wells. In most cases such a map will show that the water table is a somewhat flattened replica of the surface topography. Higher water table elevations are found under hills; lower levels are found beneath valleys. The water table may intersect the surface in the channels of streams or at the shores of lakes and marshes.

To explain why the water table assumes the basic shape of the land surface, first think of water in an open body such as a lake. Water flows into a horizontal surface because it encounters little resistance. In the subsurface, water in the saturated zone has to use available passageways to flow through the soil, but most pores are small and resist flow. When water infiltrates down from the land surface on a hill, it tends to accumulate in a mound because there is no quick and easy escape route. The water table shape under the hill is elevated with respect to the water table level at a stream, where ground water can discharge or escape.

Ground water is recharged by infiltrating water that accumulates in the saturated zone. When the saturated zone intersects the land surface, a seep or spring will be present. Construction of wells, either free flowing or pumped, provide another way for ground water to discharge.

AQUIFERS

When geotechnical engineers are consulted in the quest to find ground water, they look for the occurrence of rocks and soils that are capable of transmitting and storing appreciable quantities of water.

Ground water can occupy the openings of any type of geologic

4 · DESIGN AND CONSTRUCTION OF WATER WELLS

material. Water-bearing rocks or *aquifers* consist of either unconsolidated (loose and granular) deposits or consolidated rocks. The term *aquifer* (from the Latin for "water") describes any geological material that is capable of storing and transmitting appreciable quantities of water. The most important requirement for a body of rock to be classified as an aquifer is that it must contain interconnected openings or pores through which water can move.

Below the land surface in most areas lies a relatively permeable layer (fig. 1-1), which in turn is underlain by less permeable materials. The top layer ranges in thickness from several meters to several hundred meters. Infiltrating water in these areas will collect in the saturated zone above the less permeable layer or *aquiclude.* Sometimes the more descriptive term *confining bed* is used to refer to an aquiclude. Both terms have identical meanings pertaining to relative permeabilities (fig. 1-2).

When a confining bed is located beneath an aquifer, it prevents water from continuing its downward movement. Ground water begins to accumulate in the permeable aquifer and move laterally along the top of the confining layer. If a confining layer is located above an aquifer, the resulting ground water that begins to accumulate above it is called a *perched aquifer* (fig. 1-3).

Water from the perched water table may emerge in the side of a valley in the form of slow seepages and trickles, termed *springs* or *seeps.* The main water table intersects the stream channel and is replenished

1-1. Drillers will first encounter permeable sands and soils in most regions. (Water Well Journal Publishing Company, Columbus, Ohio)

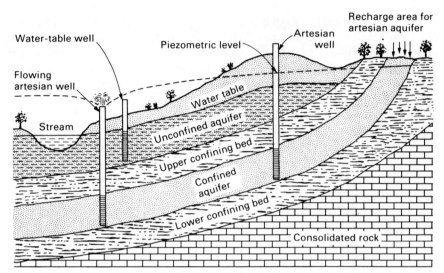

1-2. Examples of upper and lower confining beds. (After Johnson Division, 1966)

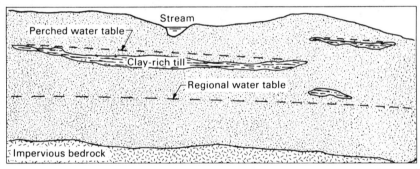

1-3. Example of a perched water table. (From Driscoll, 1986, © Johnson Division)

at more distant points where the overlying aquiclude is absent or broken. A well tapping this perched aquifer is not a reliable supply source because water replenishment for a perched water table depends on local rainfall.

Unconfined and Confined Aquifers

Water occurs in aquifers under two different conditions. When both water and air fill a permeable layer, the top of the saturated zone or water table is able to fluctuate in elevation depending on the amount of water present. The water in this aquifer is *unconfined* or said to be under water-table conditions.

The relative position of aquifers and confining beds can influence the flow of water to a water well. When a well is constructed in an aquifer that is unconfined, the standing water level elevation and the top of the saturated zone of the aquifer will be the same. This condition results because the water in an unconfined aquifer is under the same atmospheric pressure as the standing water in the well.

When water completely fills a permeable layer located under an aquiclude, the top of the saturated zone remains the same — the bottom of the confining bed. Water in this aquifer is termed *confined*, hence the term *confining bed*. Water in a confined aquifer also is referred to as being under artesian conditions, because water levels in the well respond to the pressure of the water held confined in the aquifer.

When a well is drilled into a confined or artesian aquifer, the standing water level elevation in the well will be higher than the top of the aquifer. The overlying aquiclude acts as a cap on the ground water and allows water pressure to build as more water is added from the recharge areas to the confined aquifer. Therefore, when a well is completed in a confined aquifer, the water will rise in the well to a height above the top of the aquifer because of the additional artesian pressure (see fig. 1-2). The term *artesian* was originally restricted to wells in which water actually flowed without pumping, but now the term is applied to any well in which the water rises above the elevation of the aquifer penetrated, even if the water does not reach the surface.

Artesian pressure can be mapped just like the water table. These maps are referred to as *potentiometric surface maps*. The potentiometric surface is an imaginary surface representing the level to which water will rise in wells.

ROCK TYPES

The capacity of an aquifer to store and transmit appreciable quantities of water depends on other factors besides its relative position to the surface and other rock layers. The origin and composition of the aquifer will give clues to other important aquifer properties.

All earth materials, whether consolidated or not, are collectively known as rocks. The classification of rocks on the basis of origin falls into three main categories: igneous, sedimentary, and metamorphic.

Igneous rock are formed from molten or partially molten rock, known as magma, which forms deep within the earth. When these bodies of molten rock work their way closer to the surface, they cool and solidify. An igneous rock is the hardened remains of a molten magma.

Some types of igneous rocks, such as granite, cool and solidify deep within the earth; these are referred to as *intrusive* igneous rock. *Extrusive* igneous rock is formed when the parent material, magma, is extruded or ejected onto the earth's surface. An example of an extrusive igneous rock would be lava from a volcano.

Sedimentary rocks are formed by the deposition of sediment caused by the erosion of previously existing rocks. Sediment can be deposited by water, air, or ice. At the time of deposition, most sedimentary rocks are unconsolidated (loose and granular). Over the course of time, these loose sediments will be buried and consolidated by subsequent depositional events. If this material is buried deep enough, the parent material might undergo some chemical changes as it turns into rock.

The third main group of rocks, metamorphic, are formed through the alteration of igneous or sedimentary rock by extreme heat or pressure or both. Metamorphic and igneous rocks are also referred to as crystalline rocks.

These three main rock classifications differ significantly in their capacity to store and transmit ground water. A consolidated rock body, such as a metamorphic or igneous deposit, has less pore space and fewer passageways for water than a sedimentary deposit, which can be loose or unconsolidated. For this reason, almost 60 percent of all developed aquifers consist of unconsolidated rocks.

Aquifers can be classified by parameters other than rock type. The consolidated nature of an aquifer can be used to differentiate various attributes of each type.

Unconsolidated Alluvial Aquifers

Unconsolidated sedimentary alluvial aquifers are characterized by a high percentage of permeable granular material. (*Alluvium* is a collective term used to describe the material deposited by a running water, as in stream channels or floodplains.) The unconsolidated granular material of these aquifers produces large quantities of water, most of which is recharged directly from the stream or regional rainfall.

Semiconsolidated Sedimentary Aquifers

Semiconsolidated sedimentary aquifers also are characterized by high porosity between individual grains of material, although not as high as unconsolidated alluvial aquifers. The semiconsolidated nature of this type of aquifer results from cementing agents that fill a portion of the available pore space. The cement is usually a weak bonding material such as calcium carbonate or iron oxide.

The general character of these aquifers is similar to the unconsolidated alluvial aquifers, except that semiconsolidated materials tend to be geologically older and are found at greater depths than unconsolidated aquifers.

Consolidated Sedimentary Aquifers

The consolidated sedimentary aquifers, principally of sandstone, limestone, and shale, are generally characterized by low porosity. Cementing agents fill a significant portion of the pore space, but ground water can flow along fracture planes. The quantity of water available from these aquifers depends on the volume of open pore space and the existence of fracture surfaces and cracks. Recharge to consolidated sedimentary aquifers may be local, from infiltrating rainfall, or from an overlying aquifer or regional ground-water flow from remote recharge areas.

Consolidated Crystalline Aquifers

Generally metamorphic rocks do not have any appreciable pore space that can store and transmit ground water. Some igneous rocks, such as rapidly cooled basalts, frequently develop gas bubbles, or vesicles, which provide some pore space. When these vesicles are interconnected, either by fractures, faults, or weathering processes, appreciable quantities of water can be stored. Since these aquifers are characterized by rocks with little or no pore space, the ground-water yield depends on the number of fractures present.

AQUIFER PROPERTIES

Porosity

In order to be classified as an aquifer, a body of rock must contain numerous pore spaces large enough to hold or store appreciable quantities of water. The total volume of pore space within a given volume of rock is termed *porosity* and indicates the ability of the rock or soil to hold a volume of fluid. Porosity is the ratio of pore volume to total volume, and it is usually expressed as a fraction or percentage. For example, the porosity of a clay is approximately 60 to 70 percent while a sand is about 20 to 35 percent.

Not all of the water held in the pore spaces will move through a soil or rock; some water will be retained, because of capillary pressure, as a film stuck to the walls of each pore. In rocks with very small pore spaces, such as clay, nearly all of the water is retained, even though the porosity ratio value may be high. Laboratory experiments have shown that there is a minimum size of pores (about 0.05 mm) through which water will move freely (Gilluly et al. 1959).

Porosity can be described in two ways: *specific yield*, which describes the volume of water that will drain from a rock under the influence of gravity; and *specific retention,* which describes the volume of water that will be retained in small openings as a film on the surface of larger pores (fig. 1-4).

Porosity is different in various types of rocks, depending on their origin and composition. Sedimentary rocks, such as sandstones and conglomerates, can have high porosity ratios, between 12 and 45 percent. High porosity is expected because of the relatively large openings between the rounded grains of quartz, the most common mineral grain found in sandstones.

Stream-bottom gravels and sands or beach deposits are composed of hard, well-rounded, coarse mineral grains that also have the potential for high porosity. When sands are mixed with clay, the porosity is reduced because the fine clay particles can fit into the pore spaces between the large grains, thus reducing the volume and number of the pores.

Porosity also is affected by shape, sorting and packing, and the degree of cementation of the mineral grains that make up the rock. Mineral and rock grains vary in shape from thin plates and irregular chips to nearly perfect spheres. The particular shape of a mineral grain depends on its individual crystal structure and the amount of time and energy available to allow it to develop or be destroyed.

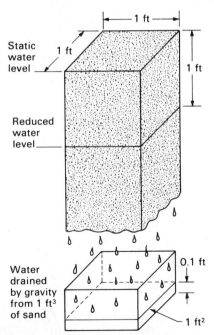

1-4. Specific yield of sand can be visualized from this diagram. Its value here is 0.1 cubic feet per cubic foot of aquifer material. (From Driscoll, 1986, © Johnson Division)

The overall uniformity of the grain size found in a rock is a function of sorting. For instance, a high-energy depositional environment such as a beach will produce a well-sorted sediment, while a low-energy depositional environment such as a tidal flat can deposit any number of sizes of sediment.

The arrangement of the grains with respect to each other is known as *packing*. As a rule, packing is loose when a sediment is first deposited, and the porosity is initially high. Compaction from the pressure of sediments deposited later progressively reduces the porosity. Cementation, which is the deposition of mineral matter in the pores surrounding the grains, can further reduce the available pore space.

Spheres of uniform size, arranged in their closest packing, have about 26 percent pore space (fig. 1-5). This is always true, regardless of whether the spheres are 1 millimeter or 5 feet in diameter. Cubic packing of spheres results in about 48 percent pore space. The shape of the grains strongly affects the tightness of packing, which in turn affects porosity values, but the presence of nonspherical grains may raise or lower the porosity, depending on how tightly the grains are packed (figs. 1-6 and 1-7).

Hydraulic Conductivity

The capacity of rock or soil to yield or transmit water to a pumping well depends on the size and amount of interconnections between the pore spaces. This property is referred to as *hydraulic conductivity*. This

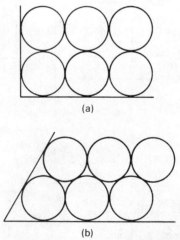

1-5. *a.* Cubic packing of spheres with a porosity of 47.65 percent.
b. Rhombohedral packing of spheres with a porosity of 25.95 percent. (From C. W. Fetter, Jr., *Applied Hydrogeology.* Westerville, Ohio: Charles E. Merrill Publishing Company, 1980. Reprinted by permission of the publisher)

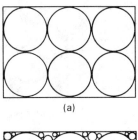

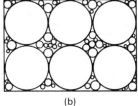

1-6. *a.* Cubic packing of spheres of equal diameter with a porosity of 47.65 percent. *b.* Cubic packing of spheres with void spaces occupied by grains of smaller diameter, resulting in a much lower overall porosity. (From C. W. Fetter, Jr., *Applied Hydrogeology.* Westerville, Ohio: Charles E. Merrill Publishing Company, 1980. Reprinted by permission of the publisher)

term replaces *permeability,* which describes the same property in the old literature.

In order to compare the water-transmitting capacity of different types of rocks, hydraulic conductivity is expressed in terms of the volume of water that would be transmitted in a unit of time through a unit cross-sectional area of rock under a unit hydraulic gradient (Heath 1984a). The metric units commonly used to express hydraulic conductivity are cubic meter for volume, square meter for area, and meter per meter for hydraulic gradient. The time frame is usually one day. The English system of measurements (inch, pound) uses cubic feet or gallons per square foot under a hydraulic gradient of a foot per foot.

To measure the water-transmitting capacity of an aquifer, the hydraulic conductivity is multiplied by the thickness of the aquifer. The aquifer thickness has to be expressed in compatible units, either meters or feet. This resulting value is called *transmissivity.*

Hydraulic conductivity values of unconsolidated, well-sorted sands and gravels are very high whereas those of dense clays and shales are very low. Hydraulic conductivity depends more on the size and geometric arrangement of the pore openings than on the percentage of pore space. For example, a gravel with 20 percent pore space is much more permeable to ground water than a clay with 35 percent pore space.

Very dense rocks, such as the igneous and metamorphic varieties, have negligible pore space in the fresh unweathered state because the mineral crystals are tightly intergrown. These rocks are, however, bro-

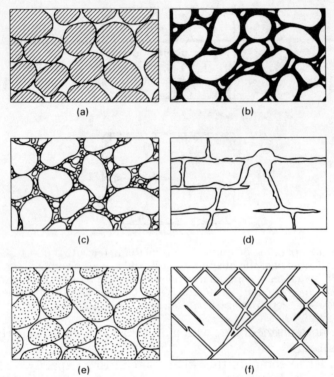

1-7. Several types of rock interstices and the relation of rock texture to porosity. *a.* Well-sorted sedimentary deposit having a high porosity. *b.* Poorly sorted sedimentary deposit having low porosity. *c.* Well-sorted sedimentary deposit consisting of pebbles that are themselves porous, so that the deposit as a whole has a very high porosity. *d.* Well-sorted sedimentary deposit whose porosity has been diminished by the deposition of mineral matter in the interstices. *e.* Rock rendered porous by solution. *f.* Rock rendered porous by fracturing. (After O. E. Meinzer, U.S. Geological Survey Water Supply Paper 489, Water Well Publishing Company, Columbus, Ohio)

ken up by numerous fractures and faults, so that the rock mass as a whole has interconnected openings through which water can move. These rocks can therefore be classified as aquifers.

Limestone, although granular as sandstone is, generally has a very low inherent porosity. Limestone transmits water not by intergranular flow but along fractures or other openings produced by weathering. Water enters these openings by precipitation or under the influence of gravity and slowly dissolves the rock, creating solution channels and enlarging the fractures to facilitate water transport. Limestone is composed primarily of calcium carbonate, and the introduction of water will dissolve it. Water falling through the atmosphere as precipitation dissolves carbon dioxide, forming a slightly acidic carbonic acid solution.

When this water enters limestone, the acid dissolves some of the calcium carbonate, thereby forming large connected channels ranging in size from a hairline fracture to a cave. In regions where limestone is near the surface and the ground water has been in contact with the limestone for long periods of time, massive solution channels have developed such as sinkholes in Florida or the Mammoth Caves in Kentucky.

Even though a rock may have a high porosity value, water cannot move freely through the mass unless the openings are interconnected and of sufficient diameter to permit flow. Therefore, the property of hydraulic conductivity, the capacity of a porous medium to transport a liquid, is of primary importance in determining the rate of groundwater movement and the amount of water (or yield) that can be drawn from a well.

EXPLORATION

The first step in developing a water supply involves studying the aerial geology. Before any specific aquifer properties can be determined, geologists use a variety of tools to assist them in understanding the hydraulic behavior of the rocks.

The most important of these tools is the geological map, which shows the distribution of various rock formations over an area and permits some understanding of the subsurface structure. Most areas of the world have been geologically mapped, although the detail and reliability of these maps vary. Hydrogeological maps, which combine information on basic geology with hydraulic behavior data and water supply capabilities, are also available for many areas.

Aerial photographs are used to prepare geological maps, and these photos can be another tool for the hydrogeologist. For example, springs or seeps often occur in areas where permeable rocks overlie impermeable material. These features are evident in aerial photos because of the changes in vegetation. Seepages also show up along faults and other major fractures, which can be detected from the air more easily than from the ground. Rocks of different permeabilities can influence the surface drainage patterns in such a way that the patterns are easily seen from the air.

Photographs from satellites are another exploration tool. These images supplement traditional aerial photographs. By using special film, certain features are enhanced and differences in soil and vegetation can be distinguished.

Photography from the air or from space is one of a group of techniques called *remote sensing*. These techniques use the electromagnetic radiation of the earth to obtain information about the earth's surface and its near surface structure. When the electromagnetic radiation being used is visible light, the image of the ground surface is an ordinary photograph.

Imagery that uses some wavelengths of infrared radiation is particularly useful in hydrogeological studies. This method, called *line scanning*, uses equipment mounted in an aircraft to build an image of the ground that depends on the amount of infrared energy that the ground surface is reflecting or emitting. This survey is taken at a time when solar heating effects are at a minimum (usually at dawn) so that the image produced is based on the temperature of the ground surface. Water has a high specific heat content, so wet areas tend to stay at a constant temperature whereas dry rock and soil temperatures fluctuate. Seepages can be identified on the infrared image. Ground water tends to stay at the same temperature throughout the year, whereas surface water is warmer in the summer and cooler in the winter. Infrared imagery can use these differences to indicate springs discharging into rivers or into the sea.

Mapping and remote sensing provide information about the ground surface. Some idea of the subsurface conditions can be inferred from this kind of information, but to find out what is below the ground surface, a borehole must be drilled. A single borehole provides information about one location only, so as a compromise between drilling a network of expensive boreholes and relying on mapping and surface imagery, geophysical exploration techniques are used.

Geophysical exploration techniques involve the measurement of various physical properties of the ground and were developed for the exploration of oil and metallic ores. Two types of geophysical methods are used: surface methods that are conducted on the land surface, and borehole methods that are conducted below the earth's surface in a well or test hole.

Surface geophysical methods measure the distance and depth to a geologic boundary. The boundary can be detected because the rocks on either side of the boundary possess different physical or electrical characteristics. These inherent characteristics are related to the rock's ability to hold and transmit ground water. Two surface geophysical techniques frequently used in ground-water investigations are *seismic* and *electrical resistivity surveys*.

Seismic surveys apply a physical shock to the earth's surface through a small exploration or heavy hammer. The shock wave is reflected and refracted by different geologic layers. The time it takes to return to the surface is recorded by geophones. This procedure is repeated at several locations.

Seismic surveys are mostly used to determine the thickness of unconsolidated deposits that overlie consolidated non-water-bearing bedrock. Since shock waves travel at different velocities in the bedrock and in unconsolidated deposits, the survey helps determine the depth to bedrock. Seismic surveys are also used to determine the depth to the water table: the velocity of shock waves is higher in saturated materials and the water table acts as a reflector. Seismic methods can identify permeability differences within an unconsolidated deposit. Once these differences are noted, a water well can be located in the area with the highest permeability.

The geophysical technique most often used to study the distribution of ground water is resistivity surveying. Since most common rock materials are poor conductors of electricity, the ability of a rock layer to conduct electricity depends entirely on the amount and conductivity of the water it contains.

In electrical resistivity surveys, four electrodes are placed in the ground along a straight line. A battery generates an electrical current that is injected into the ground through two "current" electrodes and travels through the earth. The amount of resistance of the rock layers is measured by the drop in voltage at the two "potential" electrodes. Resistivity readings are also affected by quantity of ground water.

Electrical resistivity surveys are useful in mapping buried stream channels. The survey can determine the extent and depth of the buried stream channel, indicating the most promising site for the well. Electrical resistivity surveys also may be used in coastal areas where saltwater encroaches under the land or in areas where saltwater occurs naturally. The survey will indicate areas where saltwater and freshwater meet; then a well depth that will avoid the saltwater can be determined.

For most geophysical surveys, at least one borehole is needed so that the physical measurements can be related to the geology. Geophysical measurements also are made in boreholes, and these are generally known as well-logging methods. Well or borehole logging involves lowering or raising an instrument probe into the borehole and making measurements of physical properties of the surrounding rocks or of the borehole itself.

The information gleaned from the borehole is transmitted to the surface as electrical impulses and processed in a recorder. The processed information is displayed as a log. A log is really a vertical graph of the property's depth variations that correspond to the depth below the surface. A log is interpreted by studying the shape of the curve. The shape of the trace depends on the type of log and the characteristics of the rock.

Four types of geophysical logs are used in ground-water investigations: *resistivity, spontaneous potential, gamma ray,* and *caliper.* Most logging methods (except those measuring radiation) measure a response that depends on the porosity and on the properties of the fluid filling the pores. They are very useful to the oil industry (which developed most of them) and to hydrogeologists.

Resistivity logs measure the apparent electrical resistivity of formations that are adjacent to the borehole. Generally speaking, clay has a low resistivity and plots toward the left side of the graph. Sands have a medium resistivity and plot toward the middle of the graph. Granite and limestone are examples of formations with high resistivity.

Spontaneous potential logs measure the electrical differences between the borehole fluid and the surrounding ground water. The technique is mostly used to determine the presence of saltwater. In this log, the clean (saltfree) sandstones and clays have a positive potential and plot toward the middle of the graph. The saltwater formations have a negative potential and plot toward the left side of the graph.

Gamma ray logs measure the natural gamma radiation emitted by the rocks around the borehole. Virtually all rocks contain radioactive elements, and when these elements decay, they emit energy in the form of gamma rays. In general, clays have high gamma-ray emissions and plot toward the middle of the graph. Sandstones have low counts and plot toward the left of the graph.

Caliper logs empirically measure the size of the borehole. Expanding "feeler arms" record variations in the borehole diameter as the probe is raised toward the surface. Variations in the borehole diameter indicate different rock types. Drilling through fractured limestone may produce a large hole whereas a clay layer may swell and decrease the borehole diameter.

It is difficult to determine the exact water-bearing units by using only one kind of log, so frequently a "suite" or series of logs are produced. By comparing the results of several kinds of logs, the accuracy of the determination increases.

2

Ground-Water Flow

The ability of earth materials to transmit ground water depends primarily on two factors: porosity and hydraulic conductivity. Porosity is a measure of the available pore spaces within the aquifer. Hydraulic conductivity is the ease or ability of water to flow between the pore spaces.

The local geologic conditions determine the rate and volume of ground-water flow. The rate of ground-water flow can vary between 5 feet a day (considered very fast) to 5 feet a year (considered slow). Because of such variances, it is sometimes hard to visualize the dynamics of ground-water flow.

Most of the early research on ground-water flow was done in laboratories, where flow could be empirically observed and measured. Once this pioneering work was completed, the principles were liberally applied and then modified to fit field situations. These principles are still being used today; many have been adapted for computer simulation of ground-water flow patterns.

With few exceptions, the flow of water in the ground-water environment is laminar. The flow lines in laminar flow conditions are smooth, continuous, traceable, and predictable. In contrast, surface-water flow is turbulent. Turbulent flow lines are irregular, flowing in a series of eddies and whirls (fig. 2-1), making them difficult to predict.

GROUND-WATER FLOW MECHANISMS

In laminar flow conditions, the water molecules near the center of the pore spaces move more rapidly than those closer to the walls. The water molecules near the walls of the pore space are held nearly motionless because of the molecular attraction of the walls. Water molecules in the center of the pore space follow a smooth, threadlike pattern because the resistance to flow decreases toward the center of the pore space.

Ground-water flow is ultimately derived from gravity, which draws water down through the soils to the saturated zone. As ground water accumulates in the aquifer, it begins to flow from a point of recharge to a point of discharge. Recharge areas have a higher head; discharge areas have a lower head. Topographic low areas such as streams, lakes, and

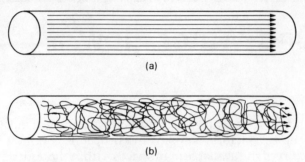

2-1. *a.* Flow paths of molecules of water in laminar flow. *b.* Flow paths of molecules of water in turbulent flow. (From C. W. Fetter, Jr., *Applied Hydrogeology*. Westerville, Ohio: Charles E. Merrill Publishing Company, 1980. Reprinted by permission of the publisher)

rivers serve as disharge points for near-surface ground water (fig. 2-2). Water wells also can serve as discharge points for ground water.

Ideally, the water table is a subdued reflection of the surface topography. The water table will rise as a horizontal plane in response to water infiltrating downward through the soil. The surface of the water table will continue to rise until it intersects the land surface. Here the water escapes from the ground and flows away on the land surface, thereby becoming surface water.

Precipitation that infiltrates through the soil from the elevated surfaces will cause the water table to mound under topographic high areas because of friction. Mounding of the water table is highest underneath the crest of a hill and is lower toward topograhic lows, creating a sloping water table. Just as surface water flows from higher head to lower head, ground water flows in response to this slope.

The slope of the water table is called the *hydraulic gradient*. The difference between a higher water table and a lower water table determines the amount of available energy that causes the ground water to flow. This difference in pressure is called the *hydraulic head*.

The hydraulic gradient can be measured by dividing the length of flow from the point of recharge to the point of discharge by the vertical distance between these two points (fig. 2-3). The hydraulic gradient is expressed as h/l, where h is the vertical distance (head) and l is the length of the flow path in which the head is measured. For example, if two wells are located 300 meters apart ($l = 300$ m), and the difference between the water table elevations is 1.3 meters ($h = 1.3$ m), the hydraulic gradient would be .0043 or 0.43 percent. This relationship between the horizontal and vertical distances forms an inclined plane. Gravity drives water down the slope of the inclined plane. The rate of groundwater flow increases as the steepness of the slope increases. Typical

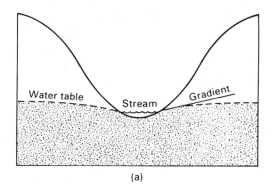

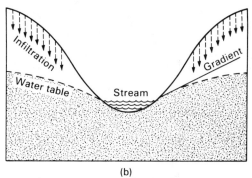

2-2. Influence of the water table gradient on baseflow. The stream in *a* is being fed by ground water with a low hydraulic gradient. A gentle rain does not produce overland flow, but infiltration raises the water table. The increased hydraulic gradient of *b* causes more baseflow to the stream, which is now deeper and has a greater discharge. (From C. W. Fetter, Jr., *Applied Hydrogeology.* Westerville, Ohio: Charles E. Merrill Publishing Company, 1980. Reprinted by permission of the publisher)

hydraulic gradients found in a ground-water environment are usually small; therefore, ground-water flow is usually slow.

An equation to express the rate of ground-water flow was originally proposed in 1856 by the French engineer Henri Darcy. While investigating the flow of water through horizontal beds of sand used for water filtration, Darcy discovered that the flow rate through a porous medium is directly proportional to the head loss and inversely proportional to the length of the flow path. This relationship is known as Darcy's law. The modern concept of ground-water flow is based on his discovery. Darcy's law may be expressed as follows:

$$Q = KA(\Delta h/\Delta l),$$

where Q is the quantity of water per unit of time, usually expressed as

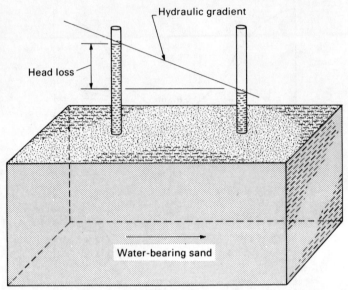

2-3. Head loss and hydraulic gradient must exist to cause flow of water through porous material. Gradient is the head loss between two points divided by the distance between the points. (© Johnson Division, 1966)

gallons per day, and K is the hydraulic conductivity of the porous media, expressed as gallons per day per square foot of aquifer. The hydraulic conductivity depends on the size, arrangement, and connections of the pore spaces within the aquifer. Other dynamic characteristics such as kinematic viscosity, density, and the effect of gravity on the fluid are held as constants if the fluid is water. These dynamic characteristics will change if the flow of other fluids (such as petroleum products) is being described. A, expressed in square feet, is the cross-sectional area through which the flow will occur and is measured at a right angle to the direction of flow. The result $\Delta h/\Delta l$ describes the hydraulic gradient.

Darcy's law states that the quantity of water (Q) flowing under laminar conditions varies directly with the hydraulic gradient ($\Delta h/\Delta l$). If the hydraulic gradient is doubled, the rate of flow through the cross-sectional area, A, also doubles. When the slope of the water table is steep, the flow of ground water is rapid.

TRANSMISSIVITY

The hydraulic conductivity, as noted earlier, defines the capacity of a *unit* volume of aquifer material to transmit water. A more practical application would be to determine the ability of the *entire* aquifer to transmit water. By modifying Darcy's original equation and incorporating aquifer variables, Darcy's law is more applicable in the field.

Transmissivity (T) is the hydraulic conductivity *(k)* multiplied by the aquifer thickness *(b)*. It is the rate of flow of ground water (expressed as gallons per day) through a vertical strip of the aquifer one foot wide, extending the full saturated thickness of the aquifer under a hydraulic gradient of 1.00 or 100 percent. Therefore:

$$T = kb.$$

Transmissivity values for aquifers range from less than 1,000 to over 1,000,000 gallons per day per foot. Aquifers with transmissivity values of less than 1,000 gpd/ft can supply only enough water for domestic wells and other low-yield uses. In general, the thickness of the aquifer is one of the major determining factors of how much water is available from the well system.

The specific capacity, or yield per unit of *drawdown* (the difference between the static water level and the pumping water level), is often used to provide a rough estimate of well efficiency. The specific capacity expressed in gallons per minute per foot (gpm/ft) is calculated by dividing the yield of a well by the drawdown, when each parameter is measured at the same time interval and discharge rate. For example, if the pumping rate is 1,000 gpm and the drawdown is 50 feet, the specific capacity of the well is 20 gpm/ft of drawdown. The specific capacity of a well will vary with the duration of pumping; in general, as pumping time and discharge increases, specific capacity decreases.

Specific capacity may be affected by the drilling and development techniques used during well construction. Well losses may occur when drilling mud invades the pore spaces of the aquifer, reducing its permeability. Improperly designed and installed well screens may cause turbulent flow near the screen intake, which also will reduce well efficiency and specific capacity.

GROUND-WATER FLOW TOWARD WELLS

When a well is idle, the head that is measured within the well is the same as the head in the aquifer adjacent to the well. The water level that stands within an idle well is known as the *static water level*.

When a well is pumped, the head inside the well decreases, and the greater head in the aquifer causes the water to flow into the well. Ground water will flow through the aquifer toward the well from all directions; this flow pattern is referred to as *converging flow*. Converg-

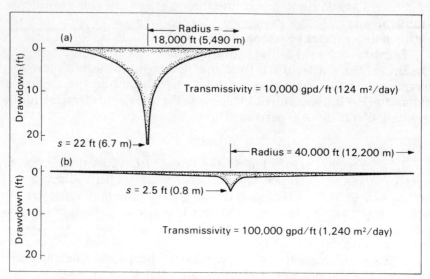

2-4. Effect of different coefficients of transmissivity on the shape, depth, and extent of the cone of depression. Pumping rate and other factors are constant. (From Driscoll, 1986, © Johnson Division)

ing flow can be represented as taking place through successive cylindrical sections that become smaller and smaller as the water approaches the well. As the area through which the ground water must flow decreases and the quantity of water remains the same, ground-water velocity increases. Accordingly, the hydraulic gradient will become steeper because of the increased velocity and decreased area through which ground water can flow. This relationship follows Darcy's law.

In an aquifer of uniform texture, the shape of the converging flow and the surface of the affected water table or piezometric surface around a pumping well is similar to an inverted cone. This shape, known as the *cone of depression,* has its apex at the water level in the pumping well and its base at the static water level in an unaffected portion of the aquifer.

The stabilized water level measured in a well that is being pumped at a constant rate is referred to as the *pumping water level.* The difference between the static water level and the pumping water level is known as the *drawdown.* Maximum drawdown occurs in the well and decreases away from the well, creating the cone of depression. At the outer limits of the cone of depression drawdown would be zero. The land surface distance from the center of the well to the outer limit of the cone of depression is known as the *radius of influence.*

The cone of depression is affected by different transmissivity values.

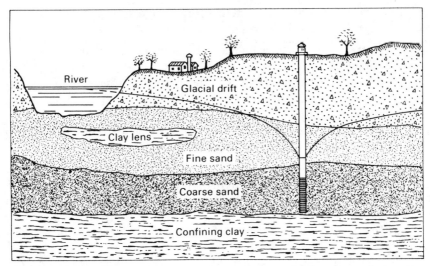

2-5. Cone of depression expanding beneath a riverbed creates a hydraulic gradient between the aquifer and the river. This can result in induced recharge to the aquifer from the river. (From Driscoll, 1986, © Johnson Division)

A low transmissivity value such as 1,000 gpd/ft will produce a deep cone of depression with steep sides and a small radius of influence. An aquifer with a high transmissivity value such as 100,000 gpd/ft will produce a shallow cone of depression. The drawdown in the well will be small and the radius of influence will have a larger areal extent (fig. 2-4).

RECHARGE AND BOUNDARY EFFECTS

When a well starts pumping, the initial quantity of water discharged from the well will be from aquifer storage immediately surrounding the well. At the onset of pumping the cone of depression is small in size. As pumping continues, the cone of depression will grow to meet the demand for ground water from aquifer storage. As the cone grows, the radius of influence increases, providing sufficient head to cause ground water to flow toward the well.

If the pumping rate is constant, then the rate of growth of the cone of depression decreases with time. Eventually the aquifer will provide enough ground water to equal the rate of pumping. This state is known as *equilibrium;* the cone of depression no longer expands or deepens. Equilibrium is reached when any of these situations occurs:

1. Natural discharge from the aquifer equals the pumping rate of the well
2. A body of surface water is intercepted by the cone of depression from which water enters the aquifer at an equivalent pumping rate
3. Recharge from precipitation and other vertical infiltration within the radius of influence equals the pumping rate
4. Recharge from adjacent aquifers equals the pumping rate

The shape of the cone of depression is symmetrical when the recharge rate is constant in all directions. However, this condition rarely occurs in nature; for example, the cone might intercept a surface stream. In this situation the cone of depression is closer to the surface near the body of water (fig. 2-5).

Conversely, the surface of the cone is depressed or lowered on the side that intercepts an impermeable boundary. Since no recharge can be obtained in that direction, the cone of depression is lowered while the other side of the cone develops a steeper slope to compensate for the lack of symmetrical recharge. Recharge areas to aquifers, such as surface water, are therefore referred to as *positive boundaries* and impermeable areas are known as *negative boundaries*.

3

Drilling Technology

Now that an identity for ground water has been established, the ways of tapping this invaluable resource need to be understood. There are two basic types of drilling: *cable tool* and *rotary*. The choice of method depends greatly on drilling conditions and depth of hole as well as access to the drilling site and availability of supplies. When drilling allows for both methods, a driller's preference should not be overlooked. Although rotary is generally considered the quicker method, an experienced cable-tool driller may drill a better constructed and better developed well than a less experienced rotary driller could.

Though cable-tool drilling has a longer history, the speed of rotary drilling indicates more applications for the future (fig. 3-1). Both methods constantly undergo technical advances as scientific research is applied to drilling dynamics. Aside from rising drilling costs, the need for potable water and the necessity of protecting invaluable ground-water sources dictate that the water-well industry cannot remain idle.

3-1. A modern control panel for a rotary rig. (Water Well Journal Publishing Company, Columbus, Ohio)

CABLE-TOOL DRILLING

History

Cable-tool drilling, also called the standard or percussion method, is the forerunner of all other drilling procedures. The first recorded use of percussion-boring tools occurred in China in about 600 B.C. for drilling brine wells a few hundred feet deep. By A.D. 1500, holes were drilled to depths of 2,000 feet, but these took generations to complete.

Many ancient wells are still in existence; in France, an artesian well drilled in A.D. 1126 is still flowing. One of the oldest wells in Europe, it is located in the province formerly known as Artois, which is also the origin of *artesian.*

Borrowing from Chinese methods, American drillers in the early nineteenth century used spring poles and similar variations to drill water and brine wells. These early drills, powered by men or horses, were replaced by the steam engine in the 1830s. Adoption of machine-powered drills gave birth to what is known as the "standard rig." Power to the four-legged derrick structure came from a stationary engine mounted apart from the rig. In 1831, drilling jars were invented, and free-falling jointed steel rods were developed during the next fifty years. Fishing tools were improved and fewer wells went unfinished.

Originally, drill-tool joints were designed with a regular straight thread, but after rechasings and continued use they were found to wear considerably, causing loose joints. Around the turn of the century, joint connections became stronger and more secure because of a new, tapered thread design. Early tool manufacturers had their own conception of the best joint, so there was no standardization of tool and joint sizes until the American Petroleum Institute was founded in 1919.

Great advances in mechanical equipment and technical skills did not occur until 1859 with the completion of the Drake oil well. Colonel E. L. Drake drilled the first commercial oil well in Titusville, Pennsylvania, using a steam-powered cable tool, which was the mainstay of the drilling industry for the next half century. During the same decade, Henry Kelly constructed a portable drilling apparatus that revolutionized the water-well industry. Soon thereafter, Kelly, Morgan and Company began commercial production of the rigs in Osage, Iowa.

By the twentieth century improvements to the steel manufacturing process allowed rigs to use steel cables for drilling instead of rope lines. Drilling tools became more effective and lasted longer.

Cable-Tool Basics

Despite the wide range of names for cable-tool drilling, the "yo-yo" method is the most descriptive. The cable-tool system is based on a simple lifting and dropping action of a string of specially designed drilling tools into the borehole. Gravity, the weight of the drill bit, and rotation from the drill string steadily break up and cut through the formations below.

Although the operation is based on simple principles, many years of experience are usually required to become proficient with the cable-tool method. A basic principle of cable-tool drilling is that the machine is always set to reach the bottom of the hole. Moreover, maintaining a tight line will maximize efficiency and reduce unnecessary damage to the cable.

Though not always the quickest method of drilling, cable tool is very economical and dependable and is likely to remain a viable method in the water-well industry. Maintenance and repair are relatively inexpensive, so with proper attention a cable-tool rig can operate effectively over a number of decades. Used primarily for water wells, cable-tool drills are ideal for drilling through unconsolidated materials such as sand and gravel, through most boulders, and through many rock formations that are fissured or cavernous.

Cable-Tool Rigs

Cable-tool rigs are manufactured in many sizes and are rated according to tool weight and depth of hole (fig. 3-2). The rig must have specifications that correspond with the type of drilling to be encountered. Because of maximum weights and stress factors, deeper holes going through rock formations will require larger, more powerful drilling rigs.

Along with determination of drilling conditions, accessibility to the drilling site should be considered. Field conditions and local road laws may restrict the use of heavier rigs. Medium-size rigs with greater portability should be considered for the drilling of a number of wells. Larger equipment that requires more preparation to move should be used at one site for longer periods of time.

The primary components of the drilling operation are known as the *tool string* (fig. 3-3), which is made up of the rope socket (attaches the tools to the cable), the drilling jars (optional), the drill stem (serves as a "guide" to the tool string), and the drill bit (does the actual cutting or boring at the hole bottom). Each component is connected by means of tool joints, and the combined weight of the tool string (ranging from about 1/2 ton to 3 tons) does the actual work. The connections are made by means of tapered, threaded pin joints that fit into the conforming box joint. The dimensions of these joints are critical to many phases of the drilling procedure. The chief measurements are the dimensions of the threaded areas, the diameters of the pin collar and box collar, and the size of the wrench squares (fig. 3-4).

Threads should be cleaned with solvent or gas and lubricated with a light grade of oil as necessary to eliminate dirt and inhibit rust. This

28 · *DESIGN AND CONSTRUCTION OF WATER WELLS*

3-2. Cable-tool rigs are manufactured in many sizes. (Water Well Journal Publishing Company, Columbus, Ohio)

ensures tighter joints and greater ease of handling when connections are made. Individual components of the tool string should be dismantled at the finish of the operation to make them last longer.

The selection of proper tools is mainly dependent on:

- well size
- capacity of the drilling rig
- height of the mast
- type of formation
- joint size

The rope or swivel socket is always the same diameter as the cable-tool joint selected. Not only does the socket make the connection between the cable and drill stem, but it allows the entire tool string to spin slightly after each blow.

The drilling jar, not used when starting a hole, is sometimes installed when drilling is underway as a precaution against the bit sticking. The jar is always positioned between the rope socket and the drill stem. The

Drilling Technology • **29**

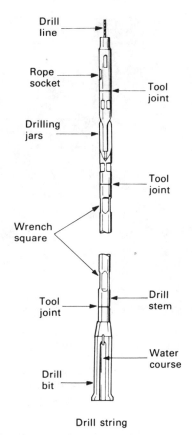

3-3. Components of the string of drill tools for cable-tool percussion drilling. (Water Well Journal Publishing Company, Columbus, Ohio)

drilling jar is actually a pair of linked, sliding steel bars. When the drill bit becomes severely stuck downhole, upward blows effectively jar it loose. The alternative would be to pull the stuck tools back steadily, which could cause the cable to snap.

As mentioned earlier, the drill stem acts as a "guide" to the rest of the tools as they are alternately lifted and dropped within the borehole walls. Most of the weight for the tool string assembly is represented by the drill stem. The longer the stem is, the heavier the weight; for water wells most are 4 or 5 feet in length.

The drill bit acts as the cutting or pulverizing agent that penetrates the formations. The bit is usually flared at the bottom and matches the width of the desired borehole. The bit chosen should have long water courses that allow more cuttings to stay suspended off the borehole floor. Then cuttings will not unnecessarily interfere with direct formation contact and deeper penetration by the bit will be achieved.

3-4. Heavy wrenches used to connect a string of drill tools. (Water Well Journal Publishing Company, Columbus, Ohio)

The cutting, turning action of the drill bit results from the elasticity and lay of the steel cable. Most cable used on cable-tool rigs is left lay; that is, the usual 6×19 or 6×21 cable is wound in a clockwise fashion. Its elasticity under tension (or when the drill bit is suspended) causes the cable to unravel partially or turn in a clockwise motion. When the bit strikes the bottom of the hole, it has turned a few degrees, making a new cut into the formation. Thus, the hole remains circular and vertical. At the same time, tension is released and the cable is allowed to recover. Left-lay cable is used because this natural turning action automatically tightens the tool joints each time the bit strikes the formation. Because of this constant tension, it is important to keep the cable well lubricated to minimize external friction and prevent internal wear.

On most rigs, the cable is led from the swivel socket over a crown sheave at the top of the derrick, then down through a sheave on the spudding beam. This beam, driven by a pitman arm or arms attached to the spudding gear, imparts the reciprocal motion to the cable and tool string. From the spudding beam, the cable passes to the bull reel on which it is wound. Kinks and sharp bends in the cable should be prevented.

Adjustments to the stroke are made with wrist pins on the spudding beam to which the pitman arm is attached. Depending on the size and make of the drilling machine, the stroke can be extended from 16 inches to as much as 48 inches; the driller varies the stroke and the speed of the gear according to drilling conditions. The drilling motion and speed should be synchronized with the gravity fall of the drilling tools and the stretch of the cable. While monitoring the drilling action, the driller must play out the right amount of cable to reach the bottom of the hole on each strike.

In cable-tool drilling, a completed borehole is the result of three major operations. First, the drill bit penetrates the various formations. Second, cuttings from the hard or soft formations must be removed periodically for the bit to make further progress. This is accomplished using a bailer, which is a section of pipe with a check valve or dart valve on the bottom end (fig. 3-5). The third major operation involves the driving of casing. Casing, used mainly in soft, alluvial formations, prevents cave-ins during the drilling operation and also helps prevent contamination from higher aquifers and formations.

The bailing operation requires that the tool string be hoisted clear of the hole, allowing the bailer to be lowered downhole. After the slurry of water and cuttings enters the bailer, the valve closes as the bailer is hoisted to the surface. Although bailing may seem like a time-consuming interruption to the actual drilling process, it plays a key role in ensuring the effectiveness of the drill bits. No regular schedule for bailing exists; frequency depends on the type of formation and the drilling speed. Gravel and heavier materials will require more bailing than clay or

3-5. The bailer is hoisted to the surface to deposit the drill cuttings. (Water Well Journal Publishing Company, Columbus, Ohio)

shale. Tendencies to use the bailer as a drill bit need to be avoided because bailers are not designed for that purpose.

While drilling through unconsolidated formations, casing is generally installed as the hole is drilled. In particularly stable conditions, casing is driven below the bit into the formation, and the resultant plug is then drilled and bailed clean. For more stable, unconsolidated formations, drilling extends 5 to 10 feet below the casing. The casing is then driven to the new bottom, and the cuttings are again bailed out.

The bottom section of casing is fitted with a heavy-walled, hardened, steel-drive shoe with a slightly larger outside diameter than the casing itself. The casing sits on a shoulder inside the drive shoe, and the lower edge of the shoe is beveled to form a cutting edge. The drive shoe prevents collapse of the casing, shaves off irregularities from the hole sides during driving, and, when driven into appropriate formations, forms a tight seal to exclude undesirable water.

As casing is driven in unconsolidated formations, vibrations cause the sides of the hole to collapse against the casing. When frictional forces become too great and the casing no longer can be driven, a smaller diameter casing is telescoped inside the first length of casing and drilling continues using a smaller drill bit. Deep holes may require four or five casing size reductions along with subsequent bit changes.

As an alternative to this frictional problem, a method developed by Melvin Church of M. Church Drilling Co., Utah, maintains hydraulic pressure on an envelope of drilling mud surrounding the driven well casing (Johnson Division 1960). Using a drive shoe slightly larger in outside diameter than the pipe, an annular space is created around the casing. A length of surface pipe larger in diameter than the pipe to be driven is installed. A rubber seal attachment is then installed at the uppermost portion of the annular space between the surface pipe and the well casing itself. Through a small diameter hole in the surface (conductor) pipe, mud is pumped into the annular space and forced downward to envelop the driven casing. The Church method not only provides lubrication for driving the casing, but also forms a seal that prevents the upward flow of water around the casing from artesian aquifers.

The only disadvantage to the Church method occurs when the pull-back method of screen emplacement is employed. The mud seal remains in place, blocking the flow of water from the aquifer. However, backwashing with water, the use of horizontal jetting techniques, and the use of polyphosphate dispersing agents will effectively remove the unwanted drilling mud.

The speed of drilling and rate of progress with cable-tool rigs depend on:

1. Hardness of the rock
2. Diameter and depth of the hole

3. Dressing of the bit
4. Weight of the tool string
5. Stroke rate and length

For penetrating hard rock formations, the drill speed is generally set at a faster rate than for drilling through clays and shales.

The capacity of a cable-tool rig depends on the weight of tools that it can handle safely; there is a limit to the depth at which a cable-tool rig can be used. On the deepest holes, added weight from the extra lengths of cable must be considered in addition to the weight of the tool string. With increased depth, more time is required in removing tools, bailing, and reinserting the tools.

Advantages

Though used less frequently, cable-tool rigs offer many advantages over rotary setups. The cost of a cable-tool rig is about one-half to two-third that of a rotary rig of equivalent capacity. Usually the rigs are relatively compact, requiring less accessory equipment than other types, and are more readily moved in rough terrain. The simplicity and ruggedness of design coupled with the ease of repair makes these rigs ideal for use in isolated areas (fig. 3-6). The low horsepower requirements are reflected in lower fuel consumption, an important consideration where fuel

3-6. Cable-tool rigs are ideal for drilling in remote areas, as found in Third World countries. (Water Well Journal Publishing Company, Columbus, Ohio)

costs are high or sources of fuel are remote. The cable-tool method also requires much less water for drilling than most other rigs.

One of the most important advantages of the cable-tool method is its ability to acquire qualitative data on the water-bearing characteristics and static heads of various formations as casing is being driven. Water-quality data can be obtained by bailer samples as each formation in turn is opened to the bottom of the casing and upper formations are cased off.

Disadvantages

Disadvantages of cable-tool rigs include the relatively slow rate of progress as well as the economical and physical limitations on depth and diameter. A further disadvantage of the cable-tool rig is the necessity of casing while drilling in unconsolidated formations.

ROTARY DRILLING

While the up-and-down motion of the cable-tool method involves a slight turning of the drill bit as it strikes the formation, the rotary drilling method extends this turning action to its maximum application. The rotary drilling process involves boring a hole by using a rapidly rotating bit to which a constant, downward force is applied.

The bit is supported and rotated by a hollow stem, composed of high-quality steel, through which a drilling fluid is circulated by means of a suitable pump. The fluid is forced through the bit openings and, in the process, cools and lubricates the cutting assembly. Because the fluid is under pressure, it naturally travels the path of least resistance up the hole-stem annulus to the surface. On its way, the fluid carries the hole cuttings in suspension up and out of the well. The cutting slurry is then diverted to a nearby mud pit where cuttings drop from suspension (fig. 3-7). The flowing mud is rerouted, via a flexible hose, down through the hollow drill stem, thus completing the cycle.

History

Although rotary drills are considered a fairly recent phenomenon, they were used by the early Egyptians in quarrying stone for the great pyramids. In 1823 water wells were drilled in Louisiana with boring tools, but the cuttings were removed by bailing.

The first serious thoughts toward mud rotary drilling came from a French engineer named Fauvelle in 1833. The idea came while he was

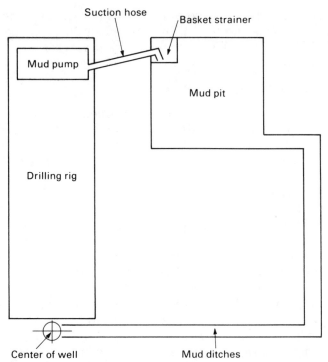

3-7. A schematic of a basic mud pit operation. (Water Well Journal Publishing Company, Columbus, Ohio)

observing a water well being drilled by a standard percussion rig. The driller's bit struck an artesian aquifer, and water spewed upward around the drill stem. Fauvelle noticed the cuttings in suspension that were being lifted up and out of the borehole.

Using his own rotary equipment with a circulating water system, Fauvelle drilled his first successful well in 1845. However, an English patent issued to Robert Beart in 1844 documents a drilling machine using a rotating tool, hollow drill rods, and circulating fluid to remove cuttings.

The revolutionary development did not significantly change the face of the drilling industry for decades. But the rotary method did become commonplace after 1901, when Captain Anthony Lucas used rotary tools to drill the Spindletop discovery well near Beaumont, Texas (UNICEF 1985).

Prior to the 1920s, a rotary rig used for water-well drilling was often called a *whirler*. The well casing itself was used as the drill pipe and was fitted with a serrated cutting shoe on the bottom edge that was slightly larger in diameter than the pipe. As the pipe was rotated, the cutting teeth loosened the formation material, and water under pressure was pumped into the casing, bringing the cuttings to the surface. This early method was only effective in soft formations free from boulders. Today the rotary method is used to drill approximately 90 percent of all wells drilled for petroleum and 80 percent of all water wells.

Components of Rotary Drilling

In mud rotary drilling, two key factors are directly related to the operation's success: the drill bit and the drilling fluid. Without the continuous circulation of drilling fluid, the bit's efficacy drops radically. The entire rig structure and each of its components directly contribute to the steady, continued operation of the bit and the drilling fluid circulation system.

Two main types of drill bits are used for rotary drilling. The lesser-used star or fishtail drag-type bit is primarily for soft, unconsolidated formations such as clay and sand (fig. 3-8). The fishtail bit has fluid courses that tend to jet the formations and therefore complement the bit's cutting action. The blades are especially effective on sticky clays.

Roller or cone-type bits are generally used for drilling through hard, consolidated formations. These bits have from two to four cone-shaped cutters mounted on roller bearings, and the intermeshing cones consist of varying lengths of cutting teeth. Bits with longer teeth are used for soft formations whereas the short, intermeshing cutters are designed for hard materials. Depending on design, the fluid courses either jet directly on the formation or wash the cutters clean for more effective cutting action.

Generally, soft formation bits are rotated faster and with a much lower bit weight. Usually, soft formation bits should be rotated at speeds from 50 to 150 revolutions per minute (RPM) with bit weights from 1,000 to 4,000 pounds per inch of bit diameter. Hard formation bits are used at speeds of 30 to 50 RPM with 2,000 to 5,000 pounds per inch of bit diameter.

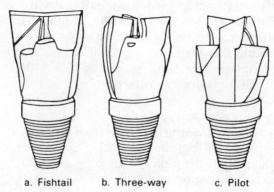

a. Fishtail   b. Three-way   c. Pilot

3-8. Types of bits. (Water Well Journal Publishing Company, Columbus, Ohio)

The drill bit operates at the lower end of the drill stem, which usually consists of three basic parts: one or more drill collars directly above the bit, one or more lengths of drill pipe, and the kelly. The drill collars are basically heavy-walled lengths of drill pipe that concentrate the desired amount of weight at the lower end of the drill stem. This weight contributes to the bit's effectiveness and helps keep the hole vertically straight.

The drill pipe, usually in 20-foot sections, is a high-strength, seamless tubing preferably made of carbon-manganese steel or a molybdenum alloy steel (fig. 3-9). Sizes for water-well drilling usually range from 2 3/8 to 4 1/2 inches and represent the outside diameter of the tubing. Proper size selection is important in order to minimize friction loss within the pipe and to lessen the power required to operate the pump. As a practical guideline for holes under 10 inches in diameter, the tool-joint outside diameter should be approximately two-thirds that of the borehole diameter.

Drill pipes are coupled together by threaded tool joints and, due to the frequent making and breaking of connections, are subject to much wear. Threads should be thoroughly cleaned and lubricated before each connection. Lubricant keeps joints from becoming too tight due to the great torsional strains. Drill pipes, making up the largest portion of the drill string, are prone to frequent failures. The principal causes of these failures are stress factors such as fatigue stress.

3-9. Drill pipe at a rotary rig job site. (Water Well Journal Publishing Company, Columbus, Ohio)

Directly above the drill pipe is the kelly, which passes through the rig-mounted rotary table (fig. 3-10). The kelly's shape—square, hexagonal, or round with lengthwise flutes cut into its outer wall—is engaged by the corresponding rotary table orifice, and thus the kelly receives its turning power.

The kelly's length is fed down through the drive bushings in the rotary table as the well is drilled deeper. Because of the kelly's work load, wall thickness is more substantial than that of drill pipe. The upper end of the kelly connects to a water swivel from which the entire drill stem hangs. Suspended from a traveling block in the derrick structure, a heavy thrust bearing between the two parts of the swivel supports the entire weight and allows the drill pipe to rotate freely.

Drilling Fluid

As a crucial factor in mud rotary methods, the consistency of drilling fluid can have significant effects on drilling time and on the actual quality of well construction and development. A good deal of study and actual field experience are helpful in perfecting the art of mud drilling. The correct use of fluids is not only important for the maintenance of affected components (for example, the drill bit and the mud pump), but

3-10. A rig-mounted rotary table. (Water Well Journal Publishing Company, Columbus, Ohio)

can determine the success or failure of the complete well system. Appropriate mud types, correct mixing, and the proper viscosity, velocity, and density of muds must be given careful attention (fig. 3-11).

Fluid Classification

The drilling fluid is classified according to the principal fluid phase as gas, water, or oil. A gas-based fluid may be (1) dry air; (2) air as a mist containing droplets of water or mud; (3) foam, that is, bubbles of air surrounded by water containing a foam-stabilizing substance; or (4) stiff foam containing film-strengthening material such as organic polymers and bentonites. Water-based fluids may contain several dissolved substances, such as alkalies, salts, and surfactants, in addition to droplets of emulsified oil and various insoluble solids carried in suspension. Oil-based drilling fluids may contain oil-soluble substances, emulsified water, and oil-insoluble materials in suspension. Oil-based muds have numerous applications in the oil industry, but are not used for drilling water wells.

Since mud composition and consistency often affect the functional and economic success of the drilling operation, proper mud

3-11. Example of a portable mud pit. (Water Well Journal Publishing Company, Columbus, Ohio)

selection should be based on these factors: the total cost of drilling; the nature of formations to be penetrated; the accessibility to supplies; the layout of the rig; the disposal of wastes; and the capabilities and limitations of the drilling equipment. Particular consideration should be paid to the skill and experience of the operator.

Drilling fluids perform four primary functions:

1. They remove cuttings that accumulate below the rotating bit.
2. They transport cuttings up the borehole to the surface where the suspended cuttings can be separated from the fluid.
3. They maintain hole stability.
4. They cool the bit.

The following additional functions will vary in importance according to local drilling conditions:

1. They prevent fluid entry from the porous rocks penetrated.
2. They reduce drilling fluid losses into permeable and loosely cemented formations.
3. They lubricate the mud pump, bit bearings, and drill string.
4. They reduce wear and corrosion of the drilling equipment.
5. They assist in collecting and interpreting information from cuttings, cores, and borehole geophysical surveys.

In the field, the term *drilling mud* can refer to any fluid ranging from muddy water, to a clay-water mixture, to a specially prepared drilling fluid consisting of measured amounts of commercial mud additives such as bentonite (fig. 3-12). In most cases the time and

3-12. A large-scale mud pit operation. (Water Well Journal Publishing Company, Columbus, Ohio)

money spent on the more effective, specially prepared drilling fluids are worth the investment.

Filter Cake

The filter cake that forms on the borehole wall represents one main reason for using drilling fluids. The somewhat flexible mud liner assists in retaining unstable, soft formations and consequently helps to prevent borehole sides from collapsing into the well. As drilling progresses, the filter cake builds up and acts as a protective layer against the eroding effects of the drilling fluid.

Although the filter cake does hold these loose formation particles in place to a certain degree, even slight pressures from surrounding geologic forces can cause the borehole wall to collapse inward. The outward hydrostatic pressure of the drilling fluid column is what prevents cave-ins from occurring. The hydrostatic pressure is the main force causing the filter cake to form in the first place. As the drilling fluid circulates through the borehole, the watery fluids tend to penetrate the crevices and porous openings of the surrounding formations. The borehole walls actually screen or filter out the particles of silt, clay, and colloids. Outward pressures cause the membrane of drilling mud to form and to remain in place.

The filter cake thickness is determined by the ability of the cake to retard water losses into the adjoining formations. The viscosity, density, and gel strength of the drilling fluid will determine the properties of the filter cake. A more permeable cake that allows drilling fluids to filter through will understandably become thicker. The ideal filter cake should effectively keep water loss to a minimum but should not penetrate formations too deeply. A thin, effective filter cake is more easily removed during the final stages of well development.

Large quantities of drilling fluids may penetrate deeply into coarse sand and gravel formations before the pores become clogged, forming a uniform seal. Problems arise when the drilling mud has penetrated water-bearing formations; the deeper the penetration, the more difficult it is to remove the mud and sand. The result may be an inferior low-yield well. High-grade bentonites (properly mixed and used in the right proportion) in drilling mud usually quicken the sealing process and keep mud invasion to a minimum.

Excessively thick filter cakes can have other adverse effects on drilling besides water losses. As the wall cake increases in thickness, it may interfere with the pulling and running of drill pipe. Also, as the borehole opening becomes progressively smaller in diameter, friction from the wall reduces the upward velocity of the drilling fluid, and cuttings cannot be removed as easily or as quickly.

Problems also may occur when the circulation system is purposely halted for installation of drill pipe or for other reasons. As soon as the mud column—with its suspended cuttings—stops its upward flow, the cuttings are drawn back to the hole bottom by gravity. Bridging of these particles may occur around tool joints and above the drill. If a mass of particles is allowed to accumulate, excessive pump pressures may be necessary to break up the accumulation and resume normal circulation. If pumping is not successful, the drilling pipe becomes stuck in the hole.

Good drilling muds can avoid the aforementioned problems by suspending the particles in place even when the pumps are temporarily shut down. When the mud flow slows or stops, properties in the mud cause gel strength to occur; that is, the drilling mud in effect gels to a certain degree and holds cuttings in place. When circulation resumes, this gelling characteristic is reversed and the drilling mud once again becomes more fluid.

As mentioned earlier, the hydrostatic pressure of the fluid prevents cave-ins and keeps the hole open for steady drilling progress. However, calculating the necessary pressure to exceed any earthen or artesian pressures in a given formation at a given depth is not a precise science. Decisions about mud weight and consistency are made by the well driller who relies on past experience. If cave-ins start to hinder drilling operations, the driller adds more bentonite or other agents to increase mud weight and therefore stabilize the walls of the hole.

Limits to mud thickness do exist, however; pump capacities must be considered. When the mud becomes too thick, not only is the pump unable to circulate the fluid effectively, but the extra load causes wear to the pump mechanism. To increase the viscosity without increasing weight, special additives such as polymers can be added. Properly mixing the right additives can increase the fluid's ability to lift cuttings out of the borehole (fig. 3-13). Generally, as the viscosity and velocity of the mud are increased, more particles are carried more quickly to the surface. Equally important is the removal of these cuttings at the settling pit before the fluid is recirculated downhole.

Besides regulating the viscosity and weight of the mud and adjusting the pump speeds, the proper layout of the mud pits is important. When suspended particles are removed, the recirculated drilling fluid is easier to pump downhole and works better to carry more cuttings in suspension. Additionally, drilling fluid free from sands and cuttings fulfills its role as a lubricant to the drill bits and other drilling components.

Control of Drilling Fluids

Though the major constituent of most drilling fluids is water, the importance of its quality is often overlooked. Water with a pH less than 7.0 may be unsatisfactory for use in mud without preliminary treatment.

3-13. Proper mixing is essential. (Water Well Journal Publishing Company, Columbus, Ohio)

For best results, the pH level of the water should be between 8.0 and 9.0 prior to the introduction of any mud additives. The on-site quality and quantity of the water may affect the performance and therefore the selection of the mud. If necessary, water can easily be treated to bring it within the proper pH levels. The pH field testing is done with chemically treated paper strips that turn color after contact with the water. The strips are then compared to a color chart to determine the pH levels.

Dissolved acidic or salty compounds in the water can seriously impair the natural properties of bentonite. Adding soda ash to acidic water before introducing the mud additives will raise the pH to desired levels. Most of the mineralized constituents of the water are removed by soda ash. Usually between 0.5 and 3.0 pounds of soda ash per 100 gallons of water is used. The treated water should then be tested for pH and calcium. If sulfides are relatively high in the make-up water, pH should be maintained above 10.0 to reduce possible premature casing and drill pipe corrosion.

As already discussed, the ability of the drilling fluid to perform its intended functions is directly affected by viscosity, mud density, gel strength, filtration characteristics, and sand content. With standard field equipment, these properties should be tested on an ongoing basis to help shorten actual drilling time and to minimize drilling costs.

Mud density is noted as pounds per gallon and is measured on a mud balance. The simply constructed mud balance consists of a calibrated arm that rests on the base. At one end of the arm is a small cup with a vented lid and at the other end is a counterweight. After the cup is filled with mud, a rider weight is adjusted at the other end of the arm and a small level indicates when perfect balance has been achieved. The weight per gallon is read directly from the calibrations on the arm next to the rider weight. The weight of the fluid helps determine the pressure placed on downhole formations. Minimum pressures are desired to prevent formation cave-ins. Weighting agents can be added as necessary, that is, when encountering high-pressure formations and it is necessary to prevent the influx of water during well construction.

Drilling fluid viscosity is checked in a simple operation using a Marsh funnel (fig. 3-14). A specified amount (1,500 cubic centimeters) of drilling mud is first poured into the plastic funnel while holding a finger over the bottom hole. The next step involves noting the amount of time it takes the fluid to drain into a measured quart cup; the fluid's viscosity is determined by the number of elapsed seconds.

As a reference point, water has a Marsh funnel viscosity of 26 seconds (and weighs 8.34 pounds per gallon). A good drilling mud weighs about 9 pounds per gallon and has a Marsh funnel viscosity ranging from 35 to 45 seconds. If the sand content of the drilling fluid

3-14. Using a Marsh funnel. (Water Well Journal Publishing Company, Columbus, Ohio)

should increase during actual drilling operations, its density also increases and the sand-laden mud will then flow through the funnel at a faster rate.

If a drilling mud with a regular Marsh funnel viscosity of 40 begins to pick up sand, it may increase its weigh to 10 pounds per gallon, but its viscosity will be decreased to about the 30-second range. Conversely, drilling muds that increase in weight as a result of drilling through and mixing with natural clay formations will have a much higher Marsh funnel viscosity than the above-mentioned 40 seconds.

Air Rotary

Air rotary drilling works on the same basic principle as mud rotary drilling except that air (that is, dry air, mist, foam, or aerated mud) instead of drilling fluid is used to remove cuttings from the borehole. More specifically, compressed air is forced downward through the hollow drill stem and then exits via the ports in the drill bit. Initially, the air jets blow away fresh cuttings and expose new rock surface to the drill bit. The jets of air then lift the cuttings off the hole bottom and force them upward. As the air stream travels up through the borehole annulus, cuttings are either carried to the surface or blown into crevices in the rock formations.

Water-well drillers began using the air rotary method after World War II, but the method has been in use since the late 1800s for operation of mining and construction equipment. A boost in the popularity of air rotary came with the first big uranium boom of the 1950s. During this time many steps also were taken in the use of air for exploratory drilling work, though all operations were not successful. The problems were thought to stem from the lack of sufficient pressure or volume. It was not unusual to find one or two large construction-type compressors (costing $25,000 to $50,000) providing air for a drilling machine costing $10,000. Extremely high operating costs forced many of the operations out of business and (fortunately for the industry) caused many individuals to reexamine the principles involved in air circulation.

After much research, the industry now has a basis for a logical, economical approach for the use of air as a circulation fluid. In many cases, air does not require high pressures when used with conventional rotary equipment. For example, most exploratory holes used for civil engineering work are less than 250 feet in depth and air pressures of 40 to 50 psi are more than adequate for recovery at such depth. Thus, the purchase of less expensive compressors can significantly reduce overall costs of the air rotary operation.

In water-well drilling, an annular velocity of 3,000 feet per minute is often considered adequate to clean the hole of cuttings when drilling with dry air, although rates ranging from 2,000 to 5,000 feet per minute have been recommended. Chart calculations are available to determine annular velocity for any given hole size, drill pipe size, and air compressor capacity.

The air rotary method has drilling applications only in consolidated formations. Most rotary drilling rigs designed to drill through these formations are equipped with a conventional mud pump in addition to the air compressor. This allows for mud drilling through unconsolidated materials before reaching bedrock. Before changing over to air rotary and to prevent possible cavings of the overburden, casing may have to be installed down to the bedrock.

There are two basic types of air drilling rigs: rotary table-drive (see fig. 3-10) and top-head drive rigs. Rotation power for the top-head drive rigs is applied from the top of the drill pipe, rather than through the rotary table mounted at the base of the rig. Most top-head drive rigs will have a single, large engine to drive the air compressor, and will either use air to provide the rotation power, or will have a hydraulic pump running off the same engine to provide hydraulic power for rotation.

Under dry drilling conditions, cuttings are removed as particles and dust. When water-bearing strata is encountered and conditions become wet, cuttings are flushed out in a mixture of air and water. When only minimal water enters the borehole, the dampened dust tends to cement small particles together into clumps or balls. These balls can become fairly large, sticking to the drill pipe and hole wall. If allowed to accumulate, these clumps can bridge across the hole. Not only does the annulus become partially or completely choked off, but the potential for stuck drill pipe also increases. To avoid such problems, small amounts of water (approximately 3 gallons per minute) should be injected into the airstream shortly after the air leaves the air compressor. This additional water keeps particles wet enough to avoid sticking and makes it easier to flush them out of the hole. Even for extremely dry hard-rock drilling conditions, injection water is useful for controlling dust.

A roller-type rock bit, similar to the kind used in mud drilling, is often used for air rotary, but types and sizes vary considerably. With certain sizes of bits, field studies have shown that both penetration rate and drill bit life are increased when compared to drilling with mud. As a significant factor, air more effectively cleans the hole bottom, avoiding the regrinding of any fresh cuttings and allowing for constant, direct contact with new formations.

Air drilling holds other advantages over mud drilling: when striking water formations, the water is quickly flushed to the surface. The driller knows exactly at what depth the formation is and can determine whether or not the aquifer has sufficient yields. With air drilling, the circulation medium is not recirculated and much time is saved in not having to construct mud pits.

Drilling with Foam

Just as with dry air, the use of aerated drilling fluids increases penetration rates when compared to mud rotary drilling. Of course, as the ratio of air to liquid decreases, borehole pressures increase and the drilling rate is somewhat reduced (but is still faster than if using drilling muds).

Foam and gel foam make possible the removal of cuttings from the hole at substantially lower velocities than is possible with dry air, and at less annular pressure than is exerted by water. Whereas air is the transport vehicle in dusting and misting, the liquid film in foam drilling is what carries the cuttings upward.

Drilling with foam is not a new development, but most drillers were not familiar with it until the late 1970s because of inadequate information. Foams are very effective for use with conventional tricone bits or downhole hammers, and besides use in the water-well industry, drilling with foam is frequently employed in minerals exploration.

The foaming agent in air drilling is a biodegradable liquid mixture of anionic surfactant, which can be added to fresh, hard, or even salty water. Just as with mud drilling, the foam is forced down through the drill pipe and out through the part in the drill bit. Returning to the surface, the slow-moving column of foam has a greater capacity for carrying the cuttings than conventional drilling fluids do. This low up-well velocity reduces well erosion and significantly lowers compressed air costs.

The advantages of air drilling with the use of a foaming agent include the following:

1. It reduces air volume and pressure requirements.
2. It increases well-cleaning capabilities and borehole stability.
3. It reduces hydrostatic head.
4. It provides a method for drilling in zones of extreme loss of circulation.
5. It suppresses dust.

Reverse-Circulation Drilling

Reverse-circulation drilling employs the same principle as that found in the conventional rotary methods, but simply reverses the flow of circulation fluid; that is, the cuttings and the fluid are sucked up through the parts in the drill bit instead of the fluid being jetted out. The drilling fluid and cuttings continue up through the drill pipe, the kelly, the swivel, and the flexible hose to the rig pump that provides the suction power. From there, the cuttings are deposited in the mud pit and the drilling fluid returns to the borehole by way of channeling and gravity. The fluid continues down the annular space to the hole bottom,

48 · *DESIGN AND CONSTRUCTION OF WATER WELLS*

picks up cuttings, and then retraces its route as suction draws the fluid and cuttings up through the drill parts (fig. 3-15). Reverse-circulation rotary drilling has limited application in the petroleum industry, but has met with considerable success in mineral exploration and ground-water industries. In mineral exploration, reverse circulation is especially useful when the quality of samples is important.

In water-well drilling, the reverse-circulation procedure has been best proven for drilling in deep sand or gravel formation areas. This method has the capability of developing a well that will produce more water per cubic foot than any other method of well construction. Reverse circulation has been used in soft, prolific rock structures with some success, but is not capable of drilling in hard-rock areas. Reverse rotary is ideal in providing large-diameter, high-capacity wells for municipalities, industry, irrigation, and other large water-use projects.

Reverse circulation offers an inexpensive method for drilling large-diameter holes in soft, unconsolidated formations. Where geologic conditions are favorable, the cost per foot of borehole increases little with increased diameter; drilling cost for a 36-inch or 40-inch hole is only

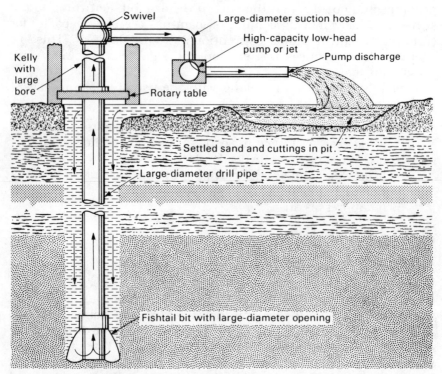

3-15. Basic principles of reverse-circulation rotary drilling. Cuttings are lifted by upflow inside drill pipe. (© Johnson Division, 1966)

moderately higher than for a 24-inch hole. Most wells drilled with the reverse-circulation method have a diameter of at least 24 inches. The diameter of the hole must be large in relation to the drill pipe so that the downward velocity of the fluid is 1 foot per second or less.

Equipment is the same as for conventional rotary, but most generally the rig is trailer-mounted due to the space requirements of the mechanical equipment used in drilling large-diameter wells. Usually a reverse rotary rig is operated around the clock until the well has reached a point of construction that ensures against the walls collapsing. To allow for operation through the night, a generator is necessary to power lights and hand tools.

Wells drilled by a reverse rotary rig can be drilled from a minimum of a 10-inch-hole diameter to a maximum of 60 inches or more under normal drilling conditions. Depths can vary from 40 feet to more than 1,500 feet. Because reverse circulation is a quick method of drilling with construction completion occurring at any time, it is necessary to have on site as much of the final material and accessory equipment as possible.

A centrifugal pump with large passageways is employed in order to handle large cuttings. One type of rig uses an ejector, operated like a large jet pump, which prevents the cuttings from passing through the centrifugal pump. Six-inch outside-diameter drill pipe is used so that rock material up to 5 inches in diameter can be brought up through the pipe. The drill pipe generally has flanged joints of approximately 11 inches in diameter. The smallest borehole that can be drilled by this method is about 18 inches in diameter, which provides sufficient annular space at the point of each flanged joint. The bit and drill pipe are rotated at speeds varying from 10 to 40 RPM.

Drill bits vary according to the type of formation materials encountered, but all have open bottoms that permit the suction and subsequent entry of drill cuttings. Some bits are mounted on an eccentric and roll on a ball-bearing race while penetrating. Other drill bits are vertical and are equipped with horizontal blades used to cut the material as they scrape through the formation. Combination pilot and drill bits are used to guide the hole progress. Larger diameter bits are mounted one on top of the other, with the uppermost bit representing the desired hole diameter.

In hard-packed materials, rock roller bits, similar to conventional rotary rock bits, are used in lieu of the blades on drag bits. The hard-packed material often snaps the blades of the drag bit, causing an undersized hole, loss of time, and repair expenses. The drilling fluid used in this method has more similarity to muddy water than to the drilling muds used in conventional rotary methods. The fine materials of silt and clay picked up in suspension from the subsurface formations are recirculated rather than settling into the mud pit. Bentonite or other drilling fluid additives are seldom added to the water for making a viscous fluid.

To prevent the hole from caving in, the fluid level is maintained at ground level. The borehole wall is supported by the hydrostatic pressures of the water column, plus the inertia of the body of water moving downward outside the drill stem. Wall erosion is not a problem because velocity in the annular space is so low.

Because additives are rarely used, some water loss occurs at each permeable layer penetrated. Some of the suspended fine particles will be filtered at the wall surface partially clogging the formation pores, but if water loss is severe, clay can be added to the water (though this should be avoided if possible). When drilling in highly permeable formations, considerable quantities of make-up water are required and must be available at all times.

Water loss can increase suddenly, and if the fluid level in the hole drops below the ground surface, caving can result. One of the major problems in reverse-circulation rotary drilling is preventing the caving of clay and shale. If the clay is wet in its native state, the adding of caustic soda to the fluid and the raising of pH to about 10.5 may be successful.

However, porous clay and shale do not stabilize with increased pH, so sodium silicate (in the ratio of 4 percent to 10 percent) may be effective. For thin-clay intervals the treatment may be made directly at the hole. If the interval is thick, then all of the fluid should be treated. If the treatment is still unsuccessful, the section should be cased or a preparation of high-viscosity, low-weight, low water-loss bentonitic muds should be used as the drilling fluid.

If casing is used, the casing diameter should not interfere with the designed completion of the well. Should the screen setting zone overlap the area that requires casing, the casing must be set to allow for setting the screens and the gravel pack material. At the appropriate time, the casing is pulled back to properly expose the screens.

Although clay sometimes helps prevent water loss by sealing off porous formations, natural clay is sometimes too effective and may cut down the yield potentials for water-bearing sand aquifers. Therefore, drilling waters should be kept as thin as possible while penetrating clay. Problems also may occur when drilling through dry clay formations; after the bit has passed through, these clay layers tend to swell from the drilling water and close in on the hole. Reaming action becomes necessary and downward progress may be delayed due to the binding and breaking of reamer blades.

Depending on the hole diameter, from 20 to 50 gallons per minute of make-up water may be needed at times when drilling through highly permeable sediments. The mud pit or water supply should have a volume of at least three times the volume of material to be removed during the drilling operation. A common circulation rate for the water used in drilling is 500 gallons per minute or more.

Drilling in coarse, dry gravel poses the greatest difficulty because of the high water-loss potential. Since high water losses usually occur above the water table, the hole should be drilled with a large auger or similar rig down to the water table or into a relatively tight formation near the water table. Surface casing is then installed and grouted prior to deepening the hole with the reverse-circulation rig.

Boulders may present problems. Cobbles larger than the drill pipe or drill bit openings cannot be brought up with the other smaller particles. When cobbles begin to collect at the hole bottom, drill pipe and bit must be pulled periodically so the stones can be grabbed and fished out by means of a steel mechanism known as an orange peel bucket. The name derives from the mechanism's similarity in appearance to an inverted, sectioned orange peel.

Dual-Tube Method

A variation of the reverse-circulation and air rotary methods of drilling is the dual-tube rotary method (also called the reverse-circulation rotary). This method combines the speed of air rotary drilling with the accuracy of reverse-circulation drilling.

The dual-tube method of drilling is based on the use of a double-walled drilling pipe that is capable of carrying air or drilling foam down to the bottom of the borehole as well as carrying cuttings out of the borehole. This double-walled drill pipe is constructed of concentric flush-jointed pipe. Air is forced down the annulus area between the two concentric pieces of drill pipe by means of a side inlet joint constructed in the uppermost section of the drilling pipe. Once at the bottom of the borehole, the air is directed toward the center of the drill pipe, released into the inner annulus area of the drilling string, allowing the air to travel back up to the surface. As the air begins to travel up the inner annulus area, material cut by the drilling bit is also carried along with it. If the cuttings are too large or if the cuttings are too heavy, foam or drilling fluid or even water can be substituted for the air.

The drill bit and drill stem are rotated by a top-head drive unit that rotates the drill pipe at the top of the drill string. Top-head drive units are usually driven by hydraulic systems. In addition to the top-head drive unit, an air compressor is needed to supply the air to lift the cuttings out of the hole. This system is preferred for obtaining precise and continuous samples of aquifers while drilling. Accurate samples can be obtained by stopping penetration of the drill bit and allowing the drill pipe to clear of cuttings. Penetration of the drill bit then continues. The cuttings that exit from the inner annulus area of the drilling pipe are an accurate representation of the drilled aquifer of soil. Sloughing of the borehole and drilling mud contaminatior are eliminated from the aquifer sample. By using air, samples are quickly and easily retrieved from the sampling exhaust by use of a cyclone separator or other similar methods.

In addition to accurate sampling, the dual-tube method of drilling provides a quick estimate of the amount of water in the aquifer. Water quality samples can be taken and analyzed while eliminating formation and aquifer damage from clay-based drilling fluids.

Disadvantages of the dual-tube method include:

- Being relatively new, the availability of equipment is somewhat limited.
- Large amounts of air are needed to drill large boreholes.
- Sloughing formations and water-saturated sands may cause the outside of the drill pipe to become stuck or bound up.

Advantages include:

- Accurate, continuous samples are easily obtained during the drilling process.
- Sample mixing and contamination of the sample from sloughing formations and erosion of the borehole wall is virtually eliminated.
- The danger of lost circulation and corresponding sample recovery is reduced.
- The borehole is clean and little or no formation damage occurs.

The dual-tube method of drilling is an excellent method to use during the exploration phase of a drilling program. The system lends itself to drilling small-diameter, highly accurate boreholes. The sampling time is greatly reduced during the drilling process. Because of its accurate sampling capabilities, this method is used quite extensively in the mineral exploration industry.

Air-Percussion Rotary Drilling

Another form of air drilling is percussion or downhole-hammer drilling. Once considered a specialty application, percussion drilling tools and techniques are gaining acceptance. This method is identical to conventional air rotary methods, except that the main source of energy for fracturing rock comes from a percussion machine connected directly to the bit. This single-cylinder, reciprocating air engine is driven by the same circulating air that is used to remove cuttings (fig. 3-16). The extremely rapid hammer blows — delivered with a weight on the bit that

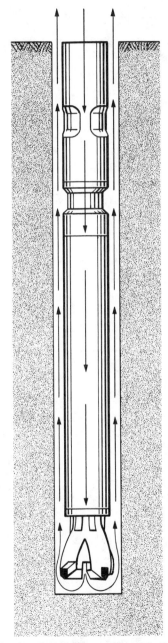

3-16. Arrows show air circulating down through the drill pipe, exiting from the drill bit, and rising up through the borehole. (Water Well Journal Publishing Company, Columbus, Ohio)

is light compared to conventional rotary weights — increase bit life, help to control hole deviation, and maintain high penetration rates.

Until the late 1960s, downhole percussion tools were designed to operate on the 100 psi compressors that were widely available to the industry. Higher-pressure tools are now being used more frequently because reliable higher-pressure compressors (200 to 250 psi) have been developed. High-pressure tools operate with approximately twice the hammer blow frequency of 100 psi tools. Thus, all the energy in the compressed air can be imparted to the piston, resulting in penetration rates double that obtained with 100 psi tools. Most carbide bit wear is caused by prolonged contact with the rock face during rotation; therefore, bit life is generally longer with high-pressure tools because, with faster penetration, less carbide is worn per foot of hole drilled.

One of the latest developments in downhole percussion air drilling is a drill bit with replaceable, tungsten-carbide buttons (fig. 3-17). The buttons used to be removed with a special punch through access holes at the back of the bit. Now the button bits have been modified so that either button setting or removal can be accomplished from the face side of the bit with a gas-pellet "gun" tool. A gas pellet either sets the buttons to a specified height above a special base and sleeve arrangement or, to replace the button, loosens the sleeve for easy removal of the entire three-piece button assembly.

3-17. Tungsten-carbide buttons. (Water Well Journal Publishing Company, Columbus, Ohio)

In percussion drilling, the selection of proper size drill pipe is critically important. Selecting a large size drill pipe can achieve equal hole-cleaning efficiency while using less air for drilling. The annular space becomes smaller and subsequently less air pressure is needed to maintain the same feet per minute velocity. Thus, hole-cleaning problems can be remedied by increasing the size of the drill pipe (from 4- to 5-inch pipe, for example) rather than by increasing compressor capacity.

Water injection into the air stream (as with conventional air rotary) is an economical method to control dust, prevent cuttings from balling up, and seal any air leakage inside the tool between closely matched parts. Water injection also cools the drilling air, ensuring that the temperature of the supplied air will not be excessive. Foam may be introduced to combat excessive ground-water influx.

Air-percussion rotary drilling is most efficient in consolidated rock formations that do not require casing. Drilling with this method in unconsolidated formations (with or without boulders) has had limited success. Prolonged drilling in wet clay may plug the air holes in the bit and stop the hammer operation. Hard and abrasive cuttings that are wet should not plug the bit, but do occasionally get into the hammer cylinder, which requires cleaning after 500 to 1,000 feet of drilling.

The depth of drilling is limited by the diameter of the hole and the volume of the compressor in use. The depth at which cuttings can be effectively removed from the hole is governed by the weight of the rock and the volume of water in the hole. Depending on the type of rock formation, the air percussion method may penetrate 50 to 100 percent faster than the standard tricone rotary bit (fig. 3-18).

In order to increase the penetration rate, the use of drill collars for weight with the drill pipe rotated on tension is recommended over pull-down pressure with the rig at ground surface. This latter procedure, as with any type of rotary drilling, places the drill pipe in compression and causes crooked holes, key seating, and so forth. Recommended pull-down pressure should be in the range of 3,500 to 4,000 pounds and can be obtained advantageously with drill collars.

At operating air pressures of 80 to 200 psi, the number of hammer percussive blows ranges from 600 to 1,000 per minute, depending on the pressure and hammer design. Rotary speeds at an optimum bit index angle of 11 degrees vary according to the bit diameter, type of formation, and hammer blows per minute. For the optimum index angle, rotary speed would vary from 18 to 30 RPM. The most economical rotating speed is that which gives the highest rate of penetration without excessive bit wear and, therefore, depends on specific field conditions.

56 · *DESIGN AND CONSTRUCTION OF WATER WELLS*

3-18. Tricone rotary bits. (Water Well Journal Publishing Company, Columbus, Ohio)

SPECIAL-APPLICATION DRILLING SYSTEMS

Other types of drilling systems (for example, jet drilling, hollow-rod drilling, auger-bucket drilling, rotary shot drilling) are designed for specialized drilling applications. Three of these methods will be discussed briefly here.

Jet Drilling

There are two variations of jet drilling, with the most basic method simply using strong jets of water to flush out earth and sands to create a moderately narrow-diameter shallow borehole. The other method uses water jets and incorporates a form of percussion drilling.

As mentioned, the simplest form of applying jetting principles involves a jetted force of water to drill shallow water wells in sandy soil. Commercial equipment is not always used; drillers may fabricate their own equipment. For example, a 1-inch piece of pipe, flattened at one end to form a 1/4-inch slot, might be used to produce the high-velocity cutting stream. Through a hose adapter, the pipe is connected to a hose, which in turn is hooked up to a hand pump or power unit.

A post-hole digger is used to start the hole to a depth of about 3 feet. The jet pipe is then positioned vertically over the hole, the water pressure is applied, and the jet stream begins cutting away at the sandy formation that is washed up and over the sides of the hole. The jet is raised and lowered, creating a churning up-and-down motion. As the jet pipe reaches deeper, the driller notes the formation materials washed to the surface to determine where the best sands or porous materials are located. The actual jetting operation usually requires only a few minutes, so the well and pipe screen assembly must be prepared in advance.

Of course, with all types of jetting, a plentiful water supply must be on hand. When the well is being jetted for a potable water supply, the water supply used for jetting should be treated with at least 50 parts per million (ppm) of chlorine to avoid possible contamination of the targeted aquifer. The most commonly used method is jet-percussion drilling, which is a quick, economical method for drilling to depths of 200 feet or more in unconsolidated soils and even some rock (like sandstone and schist). This method is similar to the cable-tool principle except that the drill rods and bit are lifted and dropped with shorter strokes. Simultaneously, water is pumped down through the drill rods to be jetted out from the drill bit, clearing the bit and also forcing the cuttings up the annulus and out of the hole. After passing through the settling pit, the water is routed to the pump for reuse (fig. 3-19).

The drill bits may vary in design, from flat chisels with jet holes on each side, to star drills with four jet holes. Usually, the first few feet are dug without water and with a blank valve over the bit openings. Then the valve is removed, circulation is started, and the drilling procedure resumes, with the drill repeatedly raised about 2 feet and then dropped. A good operator will get about 50 to 60 drops per minute.

As this procedure is taking place, the force pump is constantly circulating water through the jet holes in the drill bit. Because of residual soil and grit in the circulating water, the pump should be of the displacement type for ordinary conditions, or of the diaphragm type for extremely gritty water, and designed to provide a steady pressure on the rod line.

By using a turning clamp installed on the drill pipe, the driller turns the pipe about one-eighth of a revolution per drop until more than one-half a turn is made. This process is repeated in the other direction and then repeated back and forth until the hole is deeper than the first length of casing. This helps to ensure that the bit cuts a round hole.

Jet drilling is advantageous in larger-diameter holes of 4 to 6 inches,

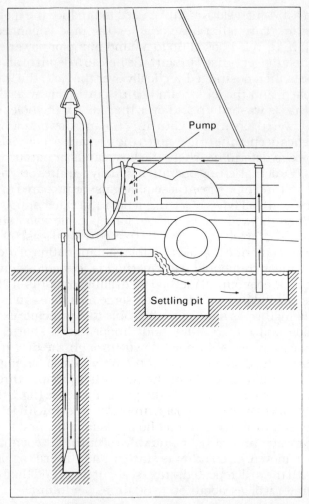

3-19. Jetting rig. (Water Well Journal Publishing Company, Columbus, Ohio)

particularly where large amounts of sand or clay are to be penetrated. When drilling through clays, a short, thin bit should be used while maintaining a stiff tension on the cable. For drilling in sand, a short, thin bit with heavy water pressure is most effective. For "quick" sands, a check valve should be installed just above the drill bit, allowing water to be pumped down through the rods, but preventing sand from being forced up through the rods. For drilling through sand with heavy gravel or boulders, a regular rock bit with sharp cutting edges is the most effective type of bit.

In unconsolidated materials, casing (fitted with a drive shoe) is normally installed as drilling progresses. To limited depths, casing may not be necessary if bentonite is added to the water and walls remain stable. If cavings do begin to occur, casing should be used, driven closely behind the bit as it progresses. In rock drilling, precautions must be taken to prevent sealing of the rock pores with mud particles from the drilling fluid. After the required depth has been reached, the hole should be flushed with clean water.

Hollow-Rod Drilling

The hollow-rod drilling method is based on the same operating principle as the cable-tool drilling method except that the cable itself does not enter the hole. The drill bit is connected to a string of hollow rods and drilling is accomplished through a vertical reciprocating motion similar to that of cable-tool drilling, but the strokes are shorter and more rapid. The drilling tool string consists of the drill bit, hollow rods, water swivel, and driving weights (fig. 3-20).

Water is maintained in the hole continuously and the drill bit is equipped with a small-ball check valve. As the drilling tools fall, fluid and cuttings are forced through this check valve which in turn prevents the cuttings from returning to the bottom. Successive strokes cause the drill pipe to fill completely and the cuttings are eventually discharged through a hose at the surface. After the cuttings are separated in the settling pit, the fluid is circulated back into the borehole.

After the well has been drilled for a few feet, casing is set in the

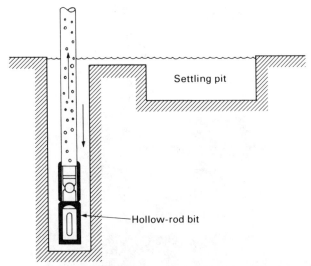

3-20. Hollow-rod drilling. (Water Well Journal Publishing Company, Columbus, Ohio)

borehole, and then drilling resumes. In unconsolidated formations, the casing follows the drill bit as drilling progresses. In consolidated formations, the drill pipe and bit are removed and the casing is driven into the formation and drilling is then resumed.

Depth of the hole and formations penetrated determine the type of operation method. Tools should be run slowly at about 35 to 40 strokes per minute until a depth of 6 to 10 feet is reached, after which the speed may be increased. For depths in excess of approximately 60 feet, the starting stroke should have an average range of 16 to 18 inches. Although the hollow-rod method can be used to depths of approximately 2000 feet, the principal drawback to this method is that it is limited to small-diameter holes.

Auger-Bucket Drilling

The auger-bucket system of drilling has been used primarily for surface-water development, or water table wells. Other economical and practical applications include gravel testing, foundation holes, pier holes, seep holes, and soil testing.

Auger-bucket drilling (or rotary bucket drilling) employs a rotary type principle to actually excavate sand and clay formations using a large-diameter auger bucket equipped with auger-type cutting blades on the bottom (fig. 3-21). The bucket is directly attached to a kelly of

3-21. Cutting blades shown at the bottom of an auger bucket. (Water Well Journal Publishing Company, Columbus, Ohio)

square configuration. This kelly passes through a kelly bar that rests across a ring gear slightly larger in diameter than the auger bucket. As power is applied to the ring gear (which acts like a rotary table) the rotating motion is transferred through the kelly bar to the kelly and then to the auger bucket. When the auger bucket is filled with formation material, the entire bucket assembly is hoisted up through the ring gear opening and is then angled away from the rig, allowing for deposit of the material on the ground.

Most kellys are designed to telescope to 75 or 100 feet in length so that drilling may proceed as long as possible before a drill-rod extension must be installed between the bucket and the kelly. As soon as one or more drill rods are used, drilling procedures are delayed as each drill rod must be dismantled to allow for clearance when hoisting the bucket. Until drill rods are used, drilling proceeds rather quickly—usually 30 to 40 feet per hour, and for some drillers, 60 feet per hour is not uncommon.

The auger-bucket system has primary applications for drilling through clay formations that remain stable without caving until pipe is installed, which then serves as casing. Drilling in sand below the water table without casing is difficult, but most caving problems are corrected by keeping the hole completely full of water. Cobbles and boulders may delay drilling operations, since the auger bucket must be removed and an orange peel bucket or other tool must be lowered into the hole to remove the obstruction.

When casing is necessary, the removal of formation materials, suction from the bucket's removal, and the weight of the concrete casing all contribute to the lowering of the casing. If caving occurs at shallow depths or the well is deep, temporary metal casing 1/4 inch to 1/2 inch thick are used. The steel casings are of telescoping diameters, similar to the ones used in caisson drilling. When satisfactory water capacity has been reached, the screen and well casing are centrally located within the drilled hole, the gravel placed, and the caissons removed.

Driven Wells

Driven wells have applications only in soft formations that are relatively free of cobbles and boulders. In favorable conditions, well points can be driven to depths of 50 feet or more. Using only hand methods, well points can be driven to depths of about 30 feet if formation conditions are favorable.

As the well point is the only area in the well unit through which water is drawn, the use of the proper point is important; many different types are available to meet varying conditions encountered in ground strata. All other factors of the well installation being normal, the well point determines the degree of success achieved with a new well (fig. 3-22).

62 • *DESIGN AND CONSTRUCTION OF WATER WELLS*

3-22. A typical well point with screen. (Water Well Journal Publishing Company, Columbus, Ohio)

To start the hole, a hand auger slightly larger than the well point is used to drill as deeply as possible. During this initial stage, care should be taken to keep the hole vertical. Riser pipe, usually 5 feet in length, is coupled to the well point. As with cable-tool couplings, tapered threads ensure stronger connections and the couplings should have tapered ends. Pipe thread compound should be applied to the pipe thread, making the joints airtight and easier to disassemble when the job is finished.

The entire riser pipe and well-point assembly is then placed vertically in the starter hole. A malleable iron drive cap is screwed to the top end of the riser pipe to absorb damage from the ensuing blows and protect the riser pipe threads (fig. 3-23). Occasionally during the operation, the drill pipe should be tightened slightly with a wrench to ensure that joints do not work loose.

Hand-driving tools include a cylindrical type of tool similar to the type used for driving steel fence posts. Depending on the heaviness of the weight that blocks off the end of the sleeve, one or two men can effectively manage this tool. A heavy maul also might be used to strike blows to the top of the drive pipe, but is not recommended because directly vertical blows are difficult to maintain.

Easier methods incorporate driving tools that are suspended from a tripod or derrick using a pulley assembly. As with any driven well operation, the driver weight, which can weigh from 75 to 300 pounds or more, should be positioned directly above the drive pipe to ensure

Drilling Technology • **63**

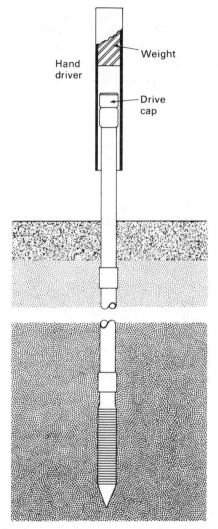

3-23. Simple tool for driving well points to depths of 15 to 30 feet. (© Johnson Division, 1966)

square blows and thus a vertical well. The heavier drive weights are quicker and more efficient in driving these wells, and light-duty cable tool rigs are often employed to handle the operation.

SAMPLING METHODOLOGY

Because sampling is an integral part of any drilling operation, the need for a discussion on sampling methodology is imperative. Although significant amounts of time and money are required of the driller in obtaining these samples and delivering them to the appropriate individual, the resultant information is invaluable not only to the driller's operation, but to the water-well industry as a whole.

The origins of most sampling techniques in the water-well industry can be traced to the petroleum industry. However, such aspects as detailed well-design features, aquifer productivity evaluations, and pump test methods have for the most part evolved from the water-well industry and its field-developed efforts since the 1960s.

Though not always in a spirit of cooperation, the spilling over of technology from one industry to the other has fortunately been a natural occurrence. Many similarities exist between crude oil and ground-water production. Both fluids characteristically flow through underground porous formations, physical laws governing their flows are identical, and wells are similarly constructed. The tapping of both resources with a minimum of drilling while gaining the maximum output greatly depends on information gained from accurate, representative samples. Reflective of the importance of sampling, much research and development in formation evaluation has occurred during the 1980s, especially in the petroleum and mining industries.

In some ways, the water-well industry holds an edge over the oil industry. Locating water is often facilitated by naturally occurring and obvious signs such as the presence of water-loving plants. Valleys are usually good indications of productive aquifers, just as surface waters (lakes, streams, and springs) are clues to the proximity and productivity of water-bearing strata.

Still, *rocks*—the term used to describe hard, consolidated material like sandstone, granite, and lava rocks as well as loose, unconsolidated materials like gravel, sand, and clay—stand as the best tool for determining the usefulness and longevity of any given aquifer. The best water carriers are gravel, sand, sandstone, and limestone, because they have the highest porosity and conductivity. Unfortunately, the bulk of rocks found in the earth's crust are clays, shales, and crystalline rocks, which offer the greatest resistivity to the flow of water.

While scientific knowledge and actual drilling experience contribute greatly to the determination of porosity and conductivity, common sense can play an equally important role. Using every resource, the geologist or hydrologist first draws up geological maps with cross-sections showing where rock formations are evident on the surface and where they are most likely to exist below the surface. In addition, any information from existing wells in the area is collected and tabulated.

A good well record will include rock samples, notations about which rock strata yield water and the amount, information from pumping or bailing tests as to the drawdown rate, and static water levels at each water-bearing strata encountered. When information is limited or no area wells exist, test wells are often needed to compile necessary data. Additional information can be found from surface indications such as quarries and exposed fault lines. From all this data the hydrologist compiles a topographic map that is similar to contour maps showing land elevations. The map shows water depths as well as the slope and direction of water flows.

Before any wells are installed, a natural balance exists that maintains a productive aquifer. The degree to which wells increase the discharge must be taken into account during the planning stages. There are two alternatives: either the natural discharge must be decreased or the recharge rate must be increased. Estimates are especially important when dealing with large-capacity wells or a large number of typically sized wells.

The gathering of proper sample amounts and the choosing of proper containers and storage procedures contribute to the success of the sampling operation. The labeling (or mislabeling) of samples can determine whether the samples are a source of invaluable information or whether they are completely useless. Care should be taken with the samples as soon as they are collected. Each sample should be clearly identified as to stratum location, sampling method, and date in order to avoid mix-ups. The protection of each sample is important because contamination can come from almost any source: other samples, drilling materials, oil, dust, water, and so forth. Samples should be handled gently to avoid excess disturbance and the breaking up of larger pieces from which much data can be collected. In some instances, samples should be protected from drying out so that they do not lose their full value for geologic studies such as geotechnical foundation analysis.

The driller or person collecting the samples must be aware of the origin of each sample by knowing the exact depth of the bit, as well as the length of time the bit was at that particular depth. Allowances must be made for the lag time of the particles in the returning fluid or on the flights of an auger drill. For example, pump pressures and varying velocities of drilling mud can determine how long particles take to reach the surface. Also, the late arrival at the surface of coarse particles (or the early arrival of fine particles from the next interval) should be noted.

The splitting of samples is sometimes necessary. For example, half of a sample may be studied in the field or sent into the laboratory to be analyzed, and the other half might be stored for further reference. A sample is occasionally split or pared down to a more usable size. This splitting allows for various tests to be run on each representative section of the whole sample. Depending on the information desired (for example, formation identification), the collection of only the larger cuttings may be necessary. Separation of these coarser particles is easily accomplished through the use of settling pits or hand sieves. However, when a full sample (including fine particles) is required, more involved procedures become necessary. Hence, the actual value of obtaining the sample should be weighed against the costly procedure of separating samples from the drilling fluid.

In cases where the full sample is required, often the amount gathered is larger than is necessary for purposes of analyses, so the sample can be split. But the fraction thereof must be truly representative of all the particles gathered in that particular run; the total sample must be thoroughly mixed to be uniform throughout. Sample splitters (using hopper and dividing partitions) are available or the procedure can be done by hand. In this latter method, the sample is mixed and then formed into a flat cake, which is divided into halves and then quarter sections until the desired amount is obtained (fig. 3-24). Each section is always mixed before further divisions are taken.

A common method of storing cutting samples is by placing them in a sample bag as soon as they are collected. Attached to the bag should be a tag noting the project name or hole number, the date, the depth interval, and the sampling method. The bags should be organized as to depth and placed in some sort of storage container. Samples that need protection from drying out should be enclosed in plastic bags or airtight containers.

With coring methods, undisturbed samples are generally left in the tube. The cutting end of the tube denotes which is the bottom portion of the sample. The tube is then sealed at both ends for protection. If the tube ends are identical, however, the sample should be clearly marked to show the bottom end. Disturbed core samples are usually removed from the tube and placed in airtight sample jars or plastic sleeves.

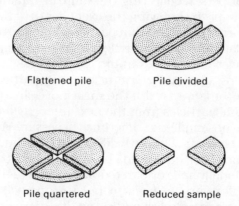

3-24. Quartering method for reducing size of sample divides a mass of material accurately. (From Driscoll, 1986, © Johnson Division)

Labeling should have the same information as that used for sample bags, including a fairly precise depth interval that denotes top to bottom measurement of the core (fig. 3-25).

All sample bags and jars should be boxed, and coring trays should be stacked for easy handling and transportation. Correct labeling of the box as to its contents is as important as the labeling of each individual sample. The containers should be protected at all times from excess vibration and from the weather.

Representative Formation Samples

Samples taken in unconsolidated formations provide necessary basic data for well design; therefore, much of the success of a completed well depends on the methods and the care exercised when obtaining formation samples. Generally, samples are taken as soon as a water-bearing formation is encountered. However, in areas where the geology is not well known, samples should be taken and stored at every lithologic change.

The choice of formation sampling method is left up to the drilling contractor. However, the contractor must collect, identify, and store representative samples in accordance with the proposed standard (see

3-25. Properly labeled core samples in plastic sleeves. (Water Well Journal Publishing Company, Columbus, Ohio)

article 45.000-000-010, page 61, of the *Manual of Water Well Construction Practices*, U.S. EPA, Office of Water Supply, 1975).

In cable-tool drilling, sample collection of the sands, gravels, and clays is accomplished through the repeated use of a bailer as drilling proceeds. The samples are usually representative of the interval drilled between bailing operations, although some contamination of the actual cuttings from holes with consolidated formations occurs from above when the cable scrapes against the uncased sections of the borehole. The bailer method can be used in both unconsolidated and consolidated formations for retrieving formation samples. In consolidated formations samples of the drill cuttings are taken immediately after the hole is bailed clean.

With cased wells in unconsolidated formations, drillers will sometimes drive the casing into the formation ahead of the drill bit rather than simply following the bit's progress. The bailer is then used to clean out the plug of material contained within the casing. Each drive may reach a depth of one to several feet depending on the formation's consistency. If plugs are too compacted to be removed by the bailer, the drill bit must be employed to loosen the material for easier bailing.

The drilling action may cause the casing to move downward by its own weight. This movement should be noted if the length of casing is used as a reference in determining the depth measurements for the samples. Also, measurement of depth by marking the drilling cable is relatively easy, but the stretching of the cable dictates that corrections be made at intervals by taking steel-line measurements.

Due to the crushing action of the drill bit in cable-tool methods, cuttings are fine. In some instances, the cuttings are ground to powder by the continued use of a dull bit; thus, it is important to keep drill bit edges sharp. Fine cuttings are a disadvantage in determining the lithology and the fossil content. During bailing operations, these fines tend to migrate toward the top while coarser materials tend to settle near the bottom of the collection sample. To ensure representative results from the sampling interval, different bailer loads should be mixed together.

When fine-grained, saturated sands and silts are involved, care must be exercised to avoid heaving of formation into the hole. This heaving can cause difficulties in sampling and logging, as it cannot be ascertained which strata the sands represent. The problem is usually solved by bailing out the material that has heaved up into the casing; the sample is then taken from the lower depth of the casing.

Cable-tool samples from granular unconsolidated formations are usually excellent. In some instances, casing may be inserted by bailing the loose formation without having to drill. The samples are obtained in a relatively unchanged state.

Various Methods

Samples of unconsolidated formations or highly unconsolidated formations are very representative if taken with a sand-pump bailer. This sand pump has a rod plunger that draws the material up through the valve and into the bailer. It should be noted that this action produces some washing of the sample, that is, more of the fines are separated from their natural placement in the formation.

A dart-valve bailer is used primarily for consolidated rock drilling, but usually requires additional drilling and breaking of gravels; hence the samples are less useful especially for sieve analysis purposes. This dart-valve bailer is not as effective for sampling sand formations. However, it can be useful if enough clay is mixed with the sands to hold them in suspension.

Drive-core sampling (or the slit-spoon method) is an effective means of obtaining accurate samples in unconsolidated strata. With this method, a tube (from 2 to 4 feet in length) is driven into the formation. The sample may be taken from a plug of formation after the casing has been driven, or it can be taken from the formation below the casing. Trial and error is used to determine where it should be taken. The core does not slip from the core barrel because the tube is overdriven. A 4-foot tube will be driven 4 1/2 to 5 feet into the formation, which compacts the material inside the tube. After the tube is brought to the surface, a few blows with a sledgehammer loosens the core, allowing it to drop out of the tube.

When drilling by direct-circulation rotary methods, samples are usually taken at regular intervals of 5 to 10 feet. This uniform system facilitates the plotting of the sample log and aids in the detection of omissions. The rock particles in the samples normally range from 1/16 to 1/2 inch, with most fragments larger than 1/4 inch. However, excessive bit weight may cause the rock particles to be so small that they cannot be used for interpretive purposes.

The return flow method of sample collection simply involves the removal of the representative sample from the discharge fluid. The drilling fluid can be run through a cutting sample box, a shale shaker, a baffle in a mud pit, or a bucket or trough that will allow the settling of particles. The method chosen depends on time, available equipment, type of formation, and driller's preference.

Because of their large size, rotary tool cuttings usually can be examined quickly under low magnification. Ordinarily, numerous core samples also are taken with rotary tools and from these cores indications of important lithologic details and even some megafossils can be obtained. In standard rotary methods, the skill and experience of the driller plays an important role in the quality of samples obtained.

Certain difficulties hinder the collection of good representative samples. For instance, rotary samples usually contain some cavings and fragments that are recirculated by the mud pump. The proportion of cavings in the samples may be large if the viscosity and circulation of the drilling mud has not been properly controlled.

As sand or sand and gravel material is transported up the borehole by the drilling fluid, some washing of the particles is inevitable. This separation can be minimized to a certain degree by controlling the drilling mud viscosity and weight. Because the finer particles are carried ahead of the coarser particles in the drilling fluid, the total sample must be recombined at the surface.

The differential settling rates of the heavy and light fragments in the mud fluid will mix the cuttings from different beds. Because the collection of samples at the surface lags behind the actual cutting of the given bed at depth, the samples usually represent a depth somewhat less than that recorded on the sample bag. This lag may amount to 20 or more feet in a 300-foot hole. The lag factor can be corrected by circulating the fluid with the drill bit suspended, so that a sufficient time interval permits the latest cuttings to reach the surface. The drill bit should be allowed to rotate during this procedure, allowing for more uniform upward flow of the fluid in the annular space. In this way, cuttings are lifted more readily from the sampling interval.

After the cuttings are collected from the settling pit, the drill bit is lowered to further penetrate the formation to a certain predetermined interval (for example, 3 feet). The drill bit is then lifted, and all these cuttings are circulated and collected before drilling begins again. Sample depths also can be corrected to some extent by timing the round-trip of some marker material (like rice) that has been circulated in the hole. From the results, a correction factor can be applied to the samples. In general, as the penetration rate increases, the lag increases. This fact can be minimized by continuing mud circulation while drill pipe is added to the drill string.

Even though a predetermined interval has been established, the action of the drill bit may indicate to the driller that a different formation interval has been encountered. At this point, drilling should be halted and samples taken as previously instructed. The new information is carefully logged before drilling begins again. Additionally, the drill bit rotation speed should be moderate, since faster rotations will not readily convey to the driller any changes in the subsurface formations.

Samples obtained during air rotary drilling are generally superior to those samples collected during mud drilling. As an example of its effectiveness, air drilling for mineral exploration has become widespread. In some types of formations, however, casing may be necessary to reduce the mixing of samples with soft, fine-grained material further up the hole.

Although the collection of reverse-circulation rotary rig samples requires experience and special equipment, a skilled operator can obtain excellent samples. The drill bit tends to merely loosen materials rather than grind them up, so the cuttings that are immediately drawn into the bit are fairly large and more representative of the formation. Also, the high-turbulent fluid velocities, usually in excess of 400 feet per minute, result in minimum lag time with little or no separation of the fines.

4

Water-Well Design

Designing a water well is not a simple engineering task. Judgment gained from experience is an important part of the design process. Geology, water quality, and the water-well drilling method are just a few of the variables over which the designer has little control. As discussed in previous chapters, the designer can obtain valuable knowledge about the potential well site through geophysical exploration and test-hole drilling prior to actual drilling. This kind of information will assist the designer during the initial phase of the design process.

Often, wells are completed by quickly conceived plans that leave little or no room for adjustment during the construction phase. Hastily installed wells may not give the customer the benefit of a long service life. The customer may be a farmer who will use the well to irrigate crops or a mining company that will use the well to dewater and stabilize the working face of the mine; the design of the water well should be matched to its application. Wells that are correctly designed for a particular purpose will provide the user with reliable service. This chapter will discuss wells that produce over 50 gallons per minute and wells that are used heavily. These wells must be designed for energy efficiency, maximum productivity, and the producton of the highest quality water possible.

THE WATER WELL

A water well is sometimes perceived as a simple structure; however, it is very complex. It can be defined as a mechanism designed for the efficient removal of ground water. In practice, the water well is designed to maximize the withdrawal of ground water while protecting the overall quality of ground water in the aquifer. After a well has been constructed, the only evidence that remains is a piece of casing that protrudes up from the ground. This is a small reminder of the complex process necessary for the design and construction of a reliable supply of ground water.

Excluding the pumping system and the distribution lines that connect the well with the surface, there are three basic parts to every well: the casing, the seal or grout, and the screen or intake portion. Wells constructed in consolidated materials (rock wells) may not have a physical screen or intake section installed because the fractured solid rock acts as the intake for the well. Rock-well design is discussed in chapter 10.

A few extra options are available on the basic model water well. A gravel pack will allow wells that are completed in fine, uniform aquifers to be more efficient and productive. Gravel packing permits control of sand particles from entering the well, which can eventually destroy the pumping equipment. Some options are not necessary for production of ground water, but they do assist in making the well more efficient to operate.

The cased portion of the well is the part that connects the aquifer with the surface. It allows access to the aquifer and provides protection against contamination from lesser quality water migrating between permeable rock layers. Casing is structurally important as it must support the sides of the borehole for the life of the well. Casing cannot be easily replaced without causing some damage to the original well design. Care in the selection and installation of the casing is crucial to the successfully completed well.

The seal or grout is the material used to seal the original borehole in the annulus outside of the casing. It serves two functions. First, the grout should seal the outside of the casing and borehole to control the movement of water vertically along the sides of the casing and borehole. Second, the material selected must be able to withstand the chemical degradation of the unwanted water as well as maintaining a seal. Ideally, a combination of materials can be used for the grout. Some grout materials are selected for their ability to add strength and protect the casing from the soil and water corrosion, while other materials are selected to resist water movement.

The heart of the well is the screen or intake section. The screen allows ground water to move freely from the aquifer into the well while stabilizing the aquifer material. Without a screen or intake section, many wells would not be able to produce enough water to be of any real use. Screens allow the well to penetrate vertically through the aquifer, taking advantage of the head and pressure situations that are found in the aquifer.

In some cases, the drilling method will determine the type of material and the design needed to construct a well. The cable-tool drilling method requires the use of steel casing. If a well is designed to have plastic casing, as for a well constructed in very corrosive soils, the cable-tool drilling method would not be the best method for constructing that well. Other drilling methods have similar limitations, and the designer will have to determine what is more important—the design features of the well or the drilling method used.

INITIAL DESIGN INFORMATION

The importance of drilling a test hole and obtaining information about the geology and aquifer characteristics of the proposed well site cannot be overemphasized. Armed with this predesign information, the designer can anticipate problems with the drilling and construction of the well before they occur. Samples of the aquifer material and water-quality data gathered from test-hole drilling are additional sources of predesign information. Based on this information, the designer can formulate the initial design of the well and begin procuring the necessary equipment on a competitive-bid basis.

Although the test-hole drilling process has been explained in chapter 3, a quick review of the process is included here. The test drilling program is divided into two phases. First is the exploratory phase of drilling, in which the designer is searching for the optimum geologic conditions where ground water will occur in usable, safe quantities. The second phase is designed to obtain as much information about the aquifer characteristics and water quality as possible. Samples of the aquifer material are obtained from the test borehole, using the best possible sampling method to ensure that the samples are of the highest quality possible. Accuracy in the sampling effort is vital for successful diagnosis and eventual well completion. In deep boreholes or in areas where the geology is complex and the exact location of the aquifer is obscured or unknown, a geophysical survey of the test hole may be necessary. Upon completion of the test hole, casing may be installed in the borehole in order to determine the quality and quantity of ground water available to the final production well.

Armed with this information, the designer can begin working on the design of the well. If preliminary information is not available, the designer may obtain this information from indirect sources. Water-quality data may be available from other wells completed in the immediate area of the proposed well. Information about the design and construction of these nearby wells also can help in the planning phase, but this information must be viewed as supplemental and not be substituted for actual data about the proposed well. Information about the yield of the aquifer also can be obtained from nearby wells, which allows the designer to estimate the yield of the final well with accuracy. Well logs from the nearby wells can provide the designer with specific data about the local drilling conditions. Zones of lost circulation and lithologic descriptions of the soils that were found at these sites will be very helpful in planning the well construction.

When information is scarce or unreliable, local drilling contractors should be interviewed. These contractors will be helpful in supplying information about local drilling conditions and aquifers. If all efforts to gain information about the proposed well site fail, the pilot hole from the production well may be used to provide some information. However, the quality of this information is suspect if delays between the pilot hole and the final production hole are long. Also, if delays between pilot and production holes are too long, the final borehole may be subjected to excessive mud invasion and collapse.

DETERMINING THE EXPECTED YIELD FROM THE WELL

One of the first steps in the design of a water well is to determine its use and expected life. The use reflects the design of the well. As an example, a well that is used to temporarily dewater or control the ground water at a construction site is not subjected to the same construction rigors as a well for municipal water supply. Although the purpose of these two wells is to pump vast quantities of water, the dewatering well will only be used for a short period of time. When the construction project is finished, the dewatering well probably will be taken out of service and abandoned. Water-quality problems such as corrosion and incrustation are of little concern to the design of the dewatering well because of its short life. The emphasis of the dewatering design is geared toward maximum production of sandfree water for the duration of the construction project.

The next step in the well design process is to determine the amount of water needed for the end user of the well. A well that is constructed to meet the present water demands of an industrial plant also could be used to accommodate future water demands (fig 4-1). In this situation, the relative merits of initially constructing a large well that can supply the future demands are weighed against constructing a series of smaller wells. Instead of building one large well, it may be prudent to phase in additional wells as water demands increase. However, the physical area to support additional wells may not exist.

Once a well is constructed, the yield is somewhat limited by the physical size and design of the whole system. There may be a temptation to increase the production of the well beyond the design capacity to produce the needed water, but by doing this, the well is subjected to serious problems such as accelerated crustation, corrosion, and growth of nonpathogenic bacteria. Irreversible damage to the well and aquifer are possible.

When the yield from the well is known, the pumping system must be reviewed. The pumping system will help determine the minimum

76 · *DESIGN AND CONSTRUCTION OF WATER WELLS*

4-1. Future water demands may necessitate a large well. (Water Well Journal Publishing Company, Columbus, Ohio)

physical size of the well structure and set the stage for the overall design features of the well. It may be tempting to choose a larger-than-needed well diameter in an attempt to eliminate the possibility of undersizing the well. But drilling a larger-than-needed well will increase the cost of drilling, labor, and equipment.

WELL DIAMETER

One long-standing myth in the water-well industry is that by doubling the diameter of the well, the quantity of water will increase appreciably. Doubling the original diameter of a well will only increase the yield by approximately 10 percent. This relationship can be seen mathematically by using a modified version of the steady-state equation for drawdown in an aquifer. The equation is rewritten in terms of specific capacity:

$$Q/s = Km/(528 \log [r0/rw]),$$

where:

> Q/s = the specific capacity of the well (gallons per minute per foot of drawdown)
> Km = the resulting transmissivity of the aquifer (permeability, K, multiplied by the screen length, m, expressed as gallons per day per foot)
> $r0$ = the radius of influence, which is the distance away from the center of the well at which the drawdown equals zero (feet)
> rw = the effective well radius, which in this case we will assume to be the same as the well radius itself (feet)

This equation can be used to compare a variety of well diameters under the same set of conditions. As an example, assume that the proposed well is 12 inches in diameter, the permeability of the aquifer is 2,500 gallons per day, and the thickness of the aquifer is 30 feet. The resulting transmissivity is 75,000 gpd/ft. The radius of influence ($r0$) is 1,450 feet. The resulting specific capacity for a well constructed as outlined above would be 41 gpm/ft of drawdown. When comparing this information to a well with a 24-inch diameter, the following changes would occur. The resulting specific capacity of the well would be 45 gpm/ft of drawdown or a 9.7 percent increase in the specific capacity (and yield) of the well. The cost of drilling the 12-inch-diameter well is substantially less than drilling and constructing the 24-inch-diameter well. (The cost of drilling these two wells is somewhat difficult to predict. Drilling costs are based on a variety of factors, which are outlined in chapter 3.)

For the sake of comparison, drilling costs would probably increase approximately 50 percent, and equipment costs (casing and screen) would increase approximately 75 percent. Another factor to consider is the availability of drilling rigs in the area that can drill and construct a 24-inch-diameter well. Procuring the necessary equipment may be more difficult and time consuming for the larger diameter well.

The axial flow of water up through the screen and casing must be examined by the designer. Too high a flow through too small a diameter screen or well casing will result in significant frictional casing and screen head loss. In addition, corrosion of the screen and casing will accelerate. Axial flow of water through the most restrictive portion of the well structure should be maintained at a velocity of less than 3 feet per second to eliminate excessive head losses. The most restrictive portion of the well structure is usually the diameter of the uppermost section of screen through which all of the water entering the well system must pass. Other restrictive portions may include liners (new sections of casing that are installed inside the old casing), packers, and seals (equipment that is used to connect the screen to the inside of the casing). Any of these restrictions must be identified and their exact physical dimensions determined. Once all of the restrictions are located, the following formula can be used to determine the velocity of water passing through the restrictions:

$$V = \frac{Q(0.3209)}{\pi r^2},$$

where:

V = velocity of water in feet per second (fps)
Q = quantity of water passing through the restriction in U.S. gallons per minute
r = radius of the restriction in inches (assuming a circular restriction)

The results of this calculation will determine if the velocity of water that flows past any of the restricted areas is greater than 3 fps. If so, the designer must determine if the yield of the system can be reduced in order to reduce the velocity. The well diameter can now be calculated, taking into account the physical restrictions of the well such as the pumping system. The final diameter of the well must be capable of (1) supporting the yield of the well without exceeding the axial velocity; (2) obtaining water from the aquifer as efficiently as possible without being oversized; and (3) supporting the pumping system.

Table 4-1 lists recommended well diameters based on the pump sizes used in the water-well industry. The designer should check with the individual manufacturer of the pumping system that is being planned for the well to determine if the well diameter is adequate. For

Table 4-1. Recommended Well Diameters Based on Pump Size.

Yield (gpm)	Optimum Casing Diameter	Minimum Casing Diameter
less than 100	6 ID	5 ID
75 to 175	8 ID	6 ID
150 to 400	10 ID	8 ID
350 to 650	12 ID	10 ID
600 to 900	14 OD	12 ID
850 to 1,300	16 OD	14 OD
1,200 to 1,800	20 OD	16 OD
1,600 to 3,000	24 OD	20 OD

SOURCE: Johnson Division 1966, p. 186.
gpm = gallons per minute
 ID = inside diameter
 OD = outside diameter

example, a small industrial plant needs a water well with a design yield of 650 gpm. The test-hole data indicates that the local aquifers are able to support the proposed well system. Based on an aquifer pumping test, a 6-inch-inside-diameter, fully penetrating well will have a specific capacity of 25 gpm/ft of drawdown. The maximum drawdown in the proposed well is 35 feet.

Table 4-1 indicates that at 650 gpm, three different well diameters can be considered. The well can be between 10 inches inside diameter to 14 inches outside diameter. The designer has checked with several manufacturers of pumping equipment and found that all of the well diameters are acceptable as far as the pumping system is concerned. The axial velocity of the water flowing through the smallest diameter is less than 3 fps. Based on the results of the aquifer pumping test of the 6-inch well, the expected specific capacity for a 10-inch-inside-diameter well is more than adequate to meet the needs of the system. Therefore, the designer selects the smallest of the three different diameters for the design of the well.

If the designer knows that the industrial plant has scheduled an expansion in the near future, which could increase the demand for water from 200 gpm to 850 gpm, the well designer may opt for a larger well diameter. A smaller-diameter well would not be capable of adequately supporting the new demand because the axial velocity would be excessive. Good design should incorporate as much information concerning expansion plans as possible.

SELECTING THE CASING TYPE

Once the diameter of the well has been determined, the next step in the design process is selection of the type of casing material. Types of casing material vary from plastic materials that are primarily used in relatively shallow, small-diameter wells to steel, which is used in the majority of well installations. Other materials include fiberglass, which is used in situations where the casing will encounter extremely corrosive waters, as in waste-injection wells. The casing material that is finally selected must be capable of withstanding the rigors of installation. It must be strong enough to support the borehole walls and pumping system once it is installed and to withstand chemical degradation and the effects of corrosion.

The casing is subjected to many physical and chemical forces. There are three physical forces present during installation: tension, column loading, and collapse pressures. The fittings and welds that connect the pieces of casing together also are subjected to these forces.

Tension is exerted on a piece of casing when it is picked up and allowed to hang in the well without support at the bottom. Tension forces are strongest at the uppermost portion of the well casing, resulting from the load or weight of the casing that is being suspended below it. As more sections of casing are added to the string of casing suspended in the borehole, more tensile forces are added to the uppermost section of casing (fig. 4-2).

If the water well is a deep installation, tensile forces can become very large and failure may occur. Tensile forces can easily be calculated beforehand, and the well casing can be designed to withstand these forces. The maximum tensile force to which casing is subjected is equal to the maximum weight of the longest string of casing that is held in suspension in the borehole. When the weight per foot of casing is known along with the length of casing that is planned for the well, the total weight of the casing can be calculated. For optimum casing design, compare the actual tensile force that is anticipated to that recommended by the manufacturer. The well designer can then adjust the casing diameter and casing wall thickness accordingly to obtain the most cost-effective design.

The fittings that connect the casing also are subject to tensile

4-2. Many factors determine casing size and thickness. (Water Well Journal Publishing Company, Columbus, Ohio)

forces. If the connection is welded, the integrity of the weld will determine the final allowable tensile loading (fig. 4-3). A formula for determining the integrity of the welds, fittings, threads, and so forth is not available due to the individual nature of each weld and each fitting that is used. Visual inspection of the on-site welds should be made to ensure integrity.

Column loading is the opposite of tensile forces. Once casing is inserted into the borehole, the casing is hung in suspension in the hole during installation. If the casing is allowed to set at the bottom of the borehole, the force exerted by that action changes the force acting on the casing. With tensile forces, we were concerned with the uppermost casing that was hanging in the well. With column loading, we are concerned with the bottom portion of the casing and screen that is resting in the hole. Here, the casing that is at the bottom of the hole must support the weight of the casing and screen above it.

Determining the column loading factor is similar to determining the tensile loading forces. The maximum column load will take place at the bottom of the casing/screen string and will be equal to the weight of the casing and screen that it is supporting. Knowing the weight of the casing and screen will determine the maximum amount of column loading to which the bottommost section of pipe is subjected.

Calculating the maximum tolerable amount of column loading that a casing or screen can tolerate is very difficult. The formulas that

4-3. A good-quality weld is required to join casing. (Water Well Journal Publishing Company, Columbus, Ohio)

are used require several assumptions to be made about the structure. The structure in the borehole is subject not only to the vertical weight of the casing above it, but also to bowing and bending. This is particularly important in deep well installations where the length of the screen and casing may be several hundred feet. The point at which bending or bowing occurs is greatly stressed. If centering guides are not placed on the casing or screen at regular intervals, the structure may rupture or bend, making the well useless. Consult the manufacturer of the screen or casing for assistance in calculating column loading and for placement of centering guides on the casing and screen.

Collapse forces can be present in a well during the installation procedure. Collapse forces are those forces that are primarily horizontal in nature and cause the casing or screen to fold inward. The primary reason for collapse forces is the difference of hydrostatic (water) head outside of the casing and inside of the casing. As the casing and screen are installed, water must pass through the screen and fill up the casing, thereby equalizing the pressures between the inside and outside of the structure. If the drilling fluid is too viscous, the fluid may not flow easily into the well casing, leaving the casing void of fluid. When this happens, there is a difference between the head of water outside of the casing and the inside of the casing. If this difference becomes too great, the screen or casing will collapse. The force can be eliminated easily by keeping the screen and casing full of liquid during the installation process. By filling the casing with fluid, differences in fluid levels are eliminated. Filling the casing with water will make the installation of plastic casing a lot easier, helping counterbalance the buoyancy factor of the plastic pipe.

Other times that collapse forces play an important part in well design are as follows: when a gravel pack is installed too quickly in a well; when the grout is poured into the annular space; during the initial phases of well development before the well and aquifer have a chance to settle and form a strong structure; when all of the free-standing water in a deep well is removed in a very short amount of time; and when a well is being pumped with a high pressure, high volume of compressed air. Again, if the collapse forces appear to be an important part in the design and construction of the well, the manufacturers of the screen and casing should be consulted. They will be able to determine the recommended pressures that the structure can tolerate, based on what is installed in the well.

Once the well is installed and the grout is set, nearly all of these forces become stabilized or eliminated. The grout and collapsed soil around the casing will help eliminate all of the tensile forces exerted on the casing. Likewise, the column loading is all but eliminated due to the friction and the support of the collapsed soils. Finally, the horizontal collapse forces are reduced because of the extra support of the grout and soils. The intake portion of a deep well can still collapse if all of the water in the well is removed at once, but this is a rare occurrence. Once the well casing is designed and determined, the intake portion of the well needs attention. The following chapter details the design of the intake portion of the well.

5

Intake Design

The intake portion of the well is the business end of every unconsolidated well. An improperly designed intake will result in an inefficient, low production rate, will cost more to operate than a properly designed well, will need more frequent maintenance, and may fail prematurely and unexpectedly.

Design of the intake portion of the well requires a combination of good engineering principles and some trial and error. No single formula or set of rules can be used to design an intake; the designer must rely in part on experience and judgment.

This chapter will explain the concepts of intake design and provide information based on experience in designing wells. It is impossible to explain every detail and nuance of well design, but with the fundamentals set down here, most designers will begin to see the logic of well design and will begin to formulate their own well design scenarios, given their area and geologic conditions.

The intake portion of the well allows ground water to be pumped from it. As a water well is pumped, it first removes water that is stored in the cased portion of the well. Then ground water must move from the aquifer into the well, passing through the openings of the intake structure, to replace the water that has been removed. Ground water is driven into the well by the head difference between the water outside the well, in the aquifer, and inside the well created by pumping the well. If the intake restricts the flow of water from the aquifer into the well, the amount of water available is reduced proportionally. Higher head differences are needed to force the same quantity of water through the intake. This basic principle is similar to that of water flow through a pipe, where high pressure or head is needed to force a known quantity of water through pipes of subsequently smaller cross-sectional areas.

An improperly designed intake will reduce not only yield, but also the efficiency of the well system. Inefficient design increases the per-unit-volume cost of obtaining water from the well. Even though the increase may seem small, a slight increase in unit-volume cost can be significant over the life of the well system. As energy costs escalate, the cost of operating a well becomes even more substantial. If the well is used as an irrigation well, the efficiency of the well system will affect the profitability of the farm operation, since the cost of operating a well becomes fixed once the well is constructed, while farm commodity prices fluctuate each season.

Because of efficiency considerations (lower pumping and energy costs), the designer may feel compelled to design the well intake more generously or always choose the intake that is known for its efficient design. Generous design may allow fine sand particles to enter the well, carried along with the ground water that is flowing into the pumped well. The sand particles will erode the pump and pumping system, adding to the maintenance cost of the well, and so adding to the total cost of the well system.

A generously designed intake can cause premature well failure. The fine sand particles that enter the well and erode the pump and pumping system may erode the intake openings. When this occurs, the intakes become enlarged and allow more sand to enter the well, thus weakening the intake further. Eventually the intake may become so weak that it collapses into the well.

A cost/benefit ratio is involved in designing a well. In some situations, the best choice for the intake may not be the most sophisticated, most expensive product on the market. Some projects, such as a construction dewatering project, may require the use of a well for only a few weeks or months. An expensive intake used in the design of these wells would have a high cost while the benefit from them would be short lived.

The intake structure plays an important part in the overall rate and amount of incrustation and chemical corrosion that occurs in the well. Poor intake design results in lower yields, inefficiency, and higher than normal head loss through the aquifer material that surrounds the intake. Higher head losses increase the potential for incrustation and corrosion to occur and will accelerate the process if the tendency to encrust or corrode is already present in the ground water. This tendency adds a hidden maintenance cost to the well system. The frequency of maintenance cannot be predicted with great accuracy until a well is in operation.

PHYSICAL PROPERTIES OF INTAKES

A variety of intake types and designs are available for use in a well. Some of these are commercially available and others are homemade. Each type and design has some beneficial uses in the ground-water industry, but not all are applicable to the design and construction of high-yielding, large-diameter water wells.

Intakes are made of a much broader range of materials than casing or pumps. For instance, intake screens are made of bronze, steel, stainless steel, Monel, or plastic. Being the working end of the well, the intake is subjected to a more intense attack of corrosion from ground water. If incrustation of the intake becomes apparent, the intake is subjected to subsequent acid treatments for the removal of the incrustation. Special materials are available for special types of wells such as those that produce water from brackish or saltwater aquifers or for wells that will be used to inject hazardous wastes into deep aquifers.

The openings of the intake can vary in their design, shape, and pattern. The openings not only allow ground water to enter the well, but they allow access to the aquifer that surrounds the well. This is particularly important during the development phase of well construction. The design of some openings keeps all aquifer material from entering the well, whereas other designs allow aquifer material to enter the well freely during development. Some openings are designed to prevent clogging. The most important factor of intake opening design is that the openings allow for the best possible access to the aquifer during development.

The perfect intake for a high-yielding water well would have the following qualities. First, it would have a high ratio of open area to closed area. Greater open area in an intake allows for good access to the aquifer during development and provides the lowest resistance to groundwater flow during pumping. The latter quality is not as important as the first, as most intakes will not cause undue head loss themselves. Numerous studies on the actual head losses of various intake designs have proven that the intake itself does not cause significant head loss in the well. However, the amount of open area does play a significant role during development and water production from an aquifer (Driscoll, Hanson, and Page 1980).

The openings of this perfect intake would be designed so that if a sand particle enters the actual opening, it does not become lodged or wedged in the opening, blocking ground-water flow. Sand particles that do enter the opening could move unrestricted through the opening and out of the well system.

In addition to the large open area, the intake should be strong enough to be capable of being set in deep wells. Strength is necessary to support both the weight of the casing above the intake as well as the forces exerted against the intake from the aquifer and ground water during installation, development, and pumping.

The perfect intake would be made of inexpensive materials that have a superior resistance to chemical corrosion from ground water. The material would also shun incrustation buildup, both biological and chemical, thus reducing the number of rehabilitation treatments. The material would also resist the chemicals used in the rehabilitation process. In addition to corrosion resistance, the intake would be manufactured of material that is a dielectric, eliminating the galvanic corrosion processes between the intake and the casing material. The perfect intake would be inexpensive, easy to obtain, and able to be quickly manufactured to the exact specifications and needs of the well and aquifer.

INTAKE TYPES

The design and manufacturing of intakes vary considerably from one manufacturer to another. Intake design types include bridge-slot and louvered-slot design; the wire-wrapped screen design; and the crude torch-cut slot method of well screening. The intakes themselves appear quite different, each having a unique quality. Each of these designs has its own beneficial uses in a water well and each design has its drawbacks.

The number and design of intake openings is perhaps the most controversial subject when discussing intakes and intake design. Sometimes opening design is the result of research, and the type, configuration, and number of openings in the intake take priority over other features of the intake. In other cases, opening design is the result of the manufacturing process or the material that is used in the construction of the intake.

Homemade Openings

The simplest of all opening designs is the saw-cut, machine-cut, or torch-cut slot. This type of intake is generally built or manufactured from a single piece of casing or pipe. Slots or openings are cut into the pipe using anything from hand saws or special diamond-tipped saws capable of accurately creating a slot of a certain dimension to cutting and welding torches. The size of the slot depends on several factors, including the type of cutting tool that is being used, the width of the cutting portion of the tool, and the conditions under which the pipe was slotted.

Torch-cut slots are the crudest type of intake opening. A welder, using a welding torch, cuts slots into a steel pipe with some kind of pattern and uniformity. Torch cutting results in very irregular openings in the pipe, both in pattern and size. Torch cutting a pipe for use as an intake is not recommended for large, high-yielding wells but may be useful in other applications, as for test wells and temporary wells.

Saw-cut openings can usually be made in the field with a small, hand-held saw. If the base pipe is plastic, a hand-held circular saw with a fine cutting blade can be used. Other types of casing materials may require specialized tools for cutting, such as diamond-tipped blades, and are generally slotted in a shop or under more controlled conditions. Slots can vary in width from 0.010 inches, using a very fine blade for soft plastic pipe, to over 0.100 inches, using a diamond-tipped blade. The length, pattern, and position of the slots is related to the cutting skills of the saw operator. Inexperienced operators tend to cut too many slots into the pipe, resulting in a significant reduction in strength.

One recurring problem with saw-cutting slots in the field are the burrs and pieces of material that become lodged inside the cut slot. Care must be taken to clean out the slots regularly so that the burrs and cut material will not block the development effort and the water flow into the well. Undercutting and overcutting pipe are other problems. Undercut pipe has too few slots, leaving an open area that is far too small to be of any use. Overcut pipe has so many slots and openings in a poorly planned pattern that the pipe has very little physical strength. When installed, the overcut pipe may collapse because of the weight of casing resting on it, or it may bend or bow in the hole. In an overcut pipe, the slots can collapse shut because of the supporting weight of casing material above it and the lack of substance to the pipe.

Machine-cut slots are cut in a manner similar to torch-cut and saw-cut slots, except that the cutting activity takes place in a controlled environment. Here, pieces of pipe are cut in a regular pattern, with special saws capable of great accuracy. Control over the length, position, and width of the slot is greatly enhanced. Slots can be machine cut into various pipe materials, including plastic, steel, and stainless steel. Machine cutting eliminates overcutting and undercutting. Most machine shops that cut pipe are aware of the limitations of the pipe material and will cut the maximum amount of slots that the material can handle without weakening the pipe's strength to an unacceptable level.

Another common type of homemade intake opening is the drilled pipe. A standard piece of pipe or casing is drilled with many small holes, creating a strainer or sieve. In coarse material, the holes are small enough to retain a portion of the aquifer material, leaving a filter of coarser material on the outside of the casing. This type of intake opening is used in domestic wells that are completed in coarse gravel deposits but is not suited for large-diameter high-yielding wells.

Punched Intake Openings

In the punched slot opening, a piece of pipe is again used as the base of the intake. A small slit is cut into the pipe. A form then pushes or punches the pipe outward, forming the opening. Bridge slots and louvered slots are the most common punched slot intakes used in the water-well industry.

The bridge slot consists of a strip of metal that is pushed out from the intake body, leaving two gaps on either side of the strip. By varying the amount that the strip is pushed out, the resulting gap can be varied. Louvered slots consist of pushing out only one edge of this strip, resulting in a louver or lip. Again, by varying the amount of pushing, the gap that results can vary accordingly. If the pipe that is used is excessively punched or punched in the wrong pattern, the integrity of the pipe may become less than desirable. Manufacturing the proper pipe in the field is nearly impossible, making the design of the pipe a function of the manufacturer.

Wire-wrapped Screen

Wire-wrapped screen is one of the most popular manufactured screens used for the completion of large-diameter, high-yielding wells (fig. 5-1). The screen offers a large amount of open area when compared to other types of screens, making the aquifer easily accessible during development and offering less resistance to flow when being pumped.

Wire-wrapped screen is manufactured by winding a shaped wire around a cylinder of evenly spaced rods. The resulting screen has the

5-1. Various types of wire-wrapped screen intakes. (Water Well Journal Publishing Company, Columbus, Ohio)

appearance of a cage with rods forming the vertical component and the wire forming the horizontal component of the screen. During manufacturing the gap between the wire can be changed, allowing for infinite combinations of the size of the openings.

Due to the nature of the wire-wrapped screen, the screen does not have a tremendous amount of strength, either vertically or horizontally. The wire and rod thickness can be changed in an effort to build strength. In cases where the intake must be strong to meet the demands of the drilling method or of the geology, a type of wire-wrapped screen construction can be accomplished on a specially made pipe that has ribs, substituting for the vertical rods. The shape of the wire also can be changed. A thinner wire shape will allow for deeper penetration of the aquifer during development as well as preventing clogging of the slots.

BASICS OF FILTER DESIGN

Two basic types of well designs are common in the ground-water industry: the gravel-packed (also called sand-packed or filter-packed well) and the naturally developed well (also referred to as a tube well). The naturally developed well uses the natural aquifer material to create a filtering zone around the intake portion of the well, keeping back the finer portion of the aquifer material, while the intake holds the filtering zone in place. This type of well is best suited for shallow, unconfined aquifers that contain a mixture of fine and coarse material.

The gravel-packed well consists of an engineered, predesigned artificial filtering pack. The "gravel" is placed near the intake portion of the well during the well construction phase and acts as a filtering media, holding back the aquifer material, while the intake holds the gravel pack in place. The term *gravel pack* refers to any filtering media that is placed in a well, and is not limited to a coarse gravel material as the name implies. Fine to medium sand is commonly used as a gravel-pack material. Gravel packing a well is usually recommended when the aquifer material consists of a fine, uniform material or when an aquifer is deep and confined.

A review of the terms used to analyze aquifer samples discussed in chapter 3 is needed to fully understand how the intake portion of a well is designed. The effective size of the aquifer sample refers to the *sieve size* in which 10 percent of the sample passes (or is finer) and 90 percent of the sample is retained (or is coarser). It is an expression of the coarseness of the sample in relationship to other samples. Fine sand has an effective size of 0.010 inches (fig. 5-2).

The *uniformity coefficient* for a sample refers to the ratio of the 90 percent retained (10 percent passing) and the 40 percent retained (60

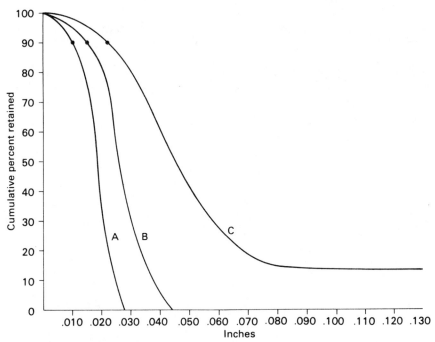

5-2. Three sand-sample analyses showing effective size. A = 0.010 inches; B = 0.015 inches; C = 0.022 inches.

percent passing) of the sample. A uniformity coefficient of 1 means that the grains of material that comprise the sample are uniform in size. A uniformity coefficient of 3.5 refers to a sample containing some fine-grained material along with some coarse material, but the majority of the sample is similar in grain size. A uniformity coefficient of 5.0 represents a sample of aquifer in which the size of the grains that make up the sample varies greatly (fig. 5-3).

DESIGNING NATURALLY DEVELOPED WELLS

Naturally developed wells are perhaps the most popular design of a water well. They require a smaller diameter borehole, construction is quicker, and the wells operate efficiently when constructed correctly. The aquifer material that is encountered is used to create the filter zone around the intake that keeps the finer portion of the aquifer stabilized, not allowing it to flow into the well. The filter zone is formed during the development phase of the well.

Thick, coarse aquifers having uniformity coefficient values greater

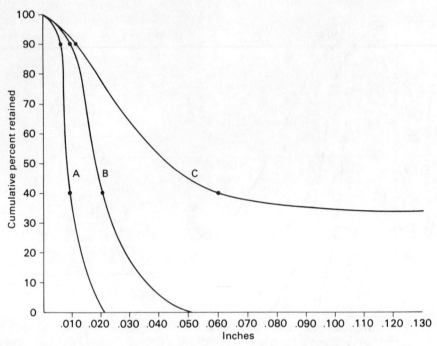

5-3. Three sand-sample analyses showing uniformity coefficient.
A = 1.5 (0.009 / 0.006); B = 2.0 (0.020 / 0.010); C = 5.0 (0.060 / 0.012).

than 3 and an effective size greater than 0.010 inches are prime candidates for being completed using a natural development design. The high uniformity coefficient value for these aquifers indicates that the material that comprises the aquifer is of such variety that it contains enough coarse grains to begin to form the filter zone that is needed to stabilize the sand. Once the coarse material begins to collect around the intake portion of the well, the remainder of the material will begin to impinge on the coarser material until, finally, the aquifer material is stabilized around the intake.

Selecting a slot size is crucial when designing the intake portion of the well. If the slots are too small they will not allow for the proper creation of the filter zone around the intake during development. Selecting a slot size that is too large will allow the coarse-grained material to be removed from the aquifer during development, reducing the amount of material that is available for creating the important zone around the intake.

Aquifers that exhibit the same characteristics and are comprised of nearly the same material throughout can be treated much differently than aquifers that are comprised of various layers or strata of material of various sizes and shapes. Thick alluvial deposits such as river deposits or ancient beach deposits generally are considered to be homogeneous in nature. Sediments such as glacial deposits or alluvial fan deposits can be much more varied in the vertical section and are considered to be heterogeneous deposits. Adjacent deposits that vary by more than two times their effective sizes can be considered heterogenous for the purposes of designing the intake. A different set of rules must be followed for the design of these aquifers.

Homogeneous Aquifers

Homogeneous aquifers that have an effective size greater than 0.010 inches, a uniformity coefficient greater than 5, and are capped with a firm layer of soil such as clay or shale, should be designed with an intake opening equivalent to the 30 percent retained (70 percent passing) size of the finest aquifer material encountered during drilling. Although the 70 percent passing may seem too liberal at first, remember that the coarse material will remain on the outside of the screen, forming a graded filter pack around the intake in order to hold back the finer portion of the aquifer material. By opening up the intake openings and allowing the natural filter to be created, wells will be more efficient and pump more water.

If the homogeneous aquifer material has an effective size greater than 0.010 inches, a uniformity coefficient greater than 5, and is not capped with a solid layer of material, the intake openings need to be selected more cautiously. An opening size equivalent to the 50 percent retained (50 percent passing) of the finest aquifer material encountered is suggested in these situations. The reason for the more conservative opening size is based on the fact that the larger opening will remove more aquifer material from the aquifer itself, thus promoting sloughing or collapse of a poorly supported aquifer.

If the sloughing or collapsing of the aquifer continues the soft sediments that are above the aquifer also will collapse, causing some minor land subsidence in the immediate area around the well and causing the finer material to work its way down to the intake portion. If this happens, the well will begin to produce or pump this material out of the well, making the water cloudy, turbid, or even sandy. The intake was originally designed for the coarser aquifer material, not the finer material.

If the homogeneous aquifer material has an effective size greater than 0.010 inches, a uniformity coefficient less than 3, and is capped with a solid layer of material, an intake opening equivalent to the 40 percent retained (60 percent passing) grain size of the finest aquifer material encountered can be used. The smaller intake opening will take into account the fact that the aquifer contains less coarse material and therefore will not build a filter zone as quickly or as effectively.

If the homogeneous aquifer material has an effective size greater than 0.010 inches, a uniformity coefficient less than 3, and is not capped with a solid layer of material, the intake openings need to be selected more cautiously. An opening size equivalent to the 60 percent retained (40 percent passing) of the finest aquifer material encountered is recommended in these situations.

Homogeneous aquifers that fall in between these guidelines can be designed either way. It will be up to the judgment and experience of the designer to select the most appropriate design in these situations. If the designer is unsure or does not have accurate samples and other information to effectively make a decision, the most conservative feasible design for the situation should be chosen.

Heterogeneous Aquifers

Aquifers that are heterogeneous in nature — those containing layers of fine and coarse material (the ratio of the effective sizes of the respective layers is greater than 2) in distinct, measurable layers — need special attention when designing the intake portion of the naturally developed well. An accurate log of the well is needed and so is analysis of the aquifer samples from each of the layers.

Once the grain-size analysis of the samples is completed, the aquifer material needs to be charted. Charting the aquifer material refers to sketching out a stratigraphic column of the aquifer and writing in the recommended intake openings in the appropriate location based on the guidelines for opening design discussed above. If the sample or the analysis of the sample is questionable, the designer must base the design of that sample on the most conservative intake opening. Once the aquifer samples are charted, the designer must use judgment to determine the exact intake opening configuration. The following rules of logic are used when designing an intake that will be placed in a heterogeneous aquifer:

1. If a finer zone overlies a coarser zone, extend the finer opening into the coarser zone no less than 2 feet.
2. If a finer zone overlies a coarser zone and the intake opening for the coarser zone is more than double that of the finer zone, extend the finer opening into the coarser zone no less than 2 feet, and increase the intake opening by 50 percent in 1-foot increments until the desired slot opening is obtained.

3. Allow for the finer intake opening to extend into the coarser zone by at least 1 foot if the coarse zone overlies the finer zone.
4. Thin gravel layers less than 3 feet thick should be ignored when designing the intake openings.

These rules take into account sloughing and collapsing of the aquifer as it is being developed and pumped. If these rules are not followed, finer aquifers will tend to collapse downward into the areas of the intake that were originally designed for the coarser zones. When this happens, the well will produce sand, encouraging more collapsing and even more sand pumping.

DESIGNING GRAVEL-PACKED WELLS

Many productive aquifers do not have an effective size that is greater than 0.010 inches. Also, many aquifers have a uniformity coefficient that is less than 3. As we discussed, coarse aquifers that are not homogeneous in nature are the best candidates for natural well completion. The more uniform aquifers cannot take advantage of larger intake openings that are acceptable for the coarser, nonuniform aquifers.

High-production wells should be gravel-packed when the following conditions exist:

1. The aquifer is fine and uniform. The effective size is less than 0.010 inches and the uniformity coefficient is less than 3.
2. The aquifer is comprised of loosely cemented sandstone. Grains of sand will continually be spalled off the rock face during pumping if the well is not gravel-packed in an effort to stabilize the formation.
3. The aquifer is very heterogeneous. That is, the aquifer contains numerous thin layers of gravel and sand in a highly laminated fashion. Gravel packing a well will reduce the concern over exact intake placement in the well during construction.
4. The well is being completed using the reverse-circulation method of drilling. Reverse circulation boreholes are very large, making naturally developed wells impractical to complete.
5. The well is being designed for maximum production with little or no maintenance. Gravel-packed wells will allow water to enter the well more freely, thereby reducing the effects of corrosion and incrustation.

6. In a thick artesian aquifer a smaller diameter screen can be used if it is gravel-packed.
7. If the aquifer is very dirty with finer sand or silt, gravel packing will help improve the well's yield by opening up the intake's slot openings and allowing for a better development of the aquifer.

Remember that gravel packing a well refers to all pregraded, designed material placed in the borehole, adjacent to the intake. The material can be as fine as a medium sand or as coarse as a cobbled gravel. Oil wells are typically gravel-packed with granular material that is finer than coffee grounds.

Gravel packs refer to a very specific type of material in the water-well industry. Gravel-pack material is typically a quartz-grained material (at least 95 percent) of a known origin (that is, the place where the gravel pack came from is not a landfill or chemical waste pit). The gravel pack is designed and sieved so that the finer material present in a gravel pack (silt and fine clay) is removed. If necessary, the gravel pack is washed several times to remove the unwanted fine material. The grains of the gravel-pack material must be well rounded. Angular grains will not work. They will tend to lock into adjacent grains, reducing the openings through which water and fine aquifer material can move.

The gravel-pack material must be highly uniform in nature. Although this condition usually does not exist in nature, there are many companies that provide a clean, quartz, well-rounded material for use as a gravel pack. Many petroleum supply houses have gravel-pack material in stock. The designer must be aware of all of the suppliers that are available to him. Although it is true that the gravel pack that is needed is not always close by, the cost of shipping the right gravel pack to the job is minimal when considering the benefits of using that gravel pack.

Gravel packs act exactly like miniature aquifers. In fact, the same attributes that make up a good gravel pack are the same as those that make up a good aquifer. A clean, well-rounded, uniform gravel has high permeability as well as good filtering ability. The high permeability nature of the gravel pack will tend to attract water to it, acting as a permeability sink in the aquifer. In addition to creating a higher permeability zone around the well intake, the gravel pack allows the designer to open the intake openings, taking full advantage of the aquifer's ability to provide water to the well without being limited by the physical nature of the intake.

The basic design of a gravel-packed well begins in the selection of a well diameter. Once the well's diameter is selected, the borehole dimensions must be calculated. Enough room in the borehole must be allotted so the proper thickness and amount of gravel pack can be placed around the intake.

Designing a gravel-packed well is a combination of selecting the correct gravel-pack thickness and the correct gravel pack itself. As the thickness of the gravel pack increases, the ratio of the gravel-pack diameter to the diameter of the aquifer also increases, but only to a limit.

If it were physically possible, a properly selected gravel pack could be placed in a 1-inch envelope between the intake and the aquifer. This envelope, although very thin, would hold back the finer aquifer material. The practical and ideal thickness of a gravel-pack envelope around a water well ranges from 4 to 6 inches. Using these thicknesses, the borehole needs to be 8 to 12 inches larger than the diameter of the intake.

As stated earlier, there is a practical limit to the thickness of the gravel pack and the ratio of the gravel pack to the aquifer material. If the gravel-pack thickness falls between the recommended thickness of 4 to 6 inches, the ratio of the mean grain size of the gravel pack as compared to the mean grain size of the aquifer material must be between 4 and 6.

If the ratio of the gravel-pack grain size to the aquifer grain size is greater than 6, serious problems can occur for the life of the well. Such gravel packs tend to become plugged with fine aquifer material that migrates toward the well, filling the pore spaces of the gravel pack. Although no sand may be produced from this practice, the pack will eventually become plugged with fine material, decreasing production and increasing the head losses around the screen. To compensate, some designers have proposed that the gravel-pack thickness be increased accordingly. Although this may work in the short term, having a larger than normal gravel-pack ratio and a very thick envelope of gravel pack around the well can cause long-term negative effects on the well's performance. The designers argue that the larger radius of the gravel pack (and theoretical radius of the well) compensates for any loss of permeability that the migration of the fine material may have on the gravel pack.

Gravel packs that have a ratio of grain-size diameters fourteen times that of the aquifer grain size have been known to allow sand particles to breach the gravel-pack filter and enter the well and well system, making the selection of these larger gravel packs unacceptable. Conversely, when gravel-pack ratios are less than 4, development is limited to such an extent that the well is never fully developed. Using a ratio of grain sizes less than 4 will reduce the intake opening to such a degree that the effort and extra expense of drilling a larger borehole are not warranted based on the benefit received.

When actually designing the gravel-pack material, the designer first begins by charting the aquifer, noting each change in aquifer material on a log. The designer then selects the finest material that is to be used in the design of the well and bases the design solely on that one formation sample. If the situation demands it, different gravel packs can be used in the same well, provided that special precautions are taken to assure that the gravel pack that is chosen for a particular depth be placed properly in the borehole. The design of this type of special well will then follow the design of a normal gravel-packed well, except that, rather than having just one design, the well will be broken into two or more separate designs.

Once the finest aquifer material is identified, the designer must then examine the adjacent zones in the aquifer, making note of those areas where the 50 percent retained size (50 percent passing size) of the aquifer material is greater than four times that of the finer material. If these coarser areas occur in the aquifer and comprise a majority of the formation, the finer zone can be sealed off by inserting a solid piece of pipe adjacent to those finer zones, taking into consideration that 3 feet of pipe should extend into the coarser zone. The intake portion that remains can be designed on the remaining coarser aquifer material.

The practice of eliminating the finer aquifer material from the design of the well will ensure that in those aquifer situations where excessively fine aquifer material causes the design of the entire gravel-packed well to become too conservative, the coarser aquifer design will prevail. When designing these types of aquifers, the designer must be careful to screen only the coarser zones. The gravel pack that will be used in this well can be calculated from the finest aquifer material that is to be screened. The 70 percent retained (30 percent passing) grain size of the finest aquifer material is multiplied by the gravel-pack ratio multipliers of 4, 5, or 6. In most cases, using a gravel-pack ratio of 5 is acceptable, except when the aquifer is comprised of fine, uniform particles or if the gravel-pack thickness is less than 4 inches around the intake.

If the aquifer material is uniform (uniformity coefficients less than 3), a uniform gravel pack should be selected. If the aquifer material is not uniform (uniformity coefficients greater than 3), the gravel-pack uniformity should follow that of the aquifer. Gravel packs should not exceed maximum uniformity coefficients of 6. Larger uniformity coefficients will tend to allow finer aquifer material into the pack and subsequently the well.

Once the initial gravel-pack curve is drawn, the designer must select the range of gravel-pack tolerances that is acceptable. Approximately 8 percent is accepted in most cases (fig. 5-4). If the aquifer is uniform or is comprised of very fine grain sizes, a tighter tolerance may be justified. The designer now has the set of specifications the gravel pack must meet. Gravel-pack material occurs naturally in areas of the country where large deposits of clean sand and gravel exist. If a gravel pack meeting the exact specifications is not found naturally, the designer may have to have the gravel pack custom mixed by the supplier. By

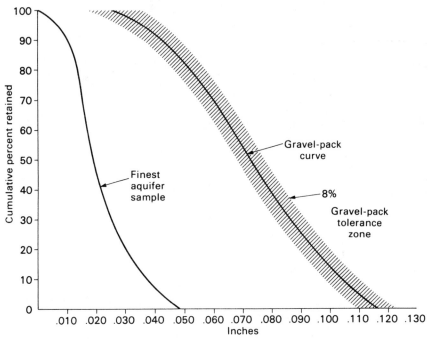

5-4. Sand analysis of the finest aquifer sample and the resulting 8 percent gravel-pack tolerance zone.

combining various mixtures of fine, medium, and coarse packs, the designer will be able to obtain a gravel pack that will meet the 8 percent specification.

Adequate distribution of grain sizes in the gravel pack is needed to assure that the gravel pack will perform its job. Mixing a coarse and a fine sand together without any medium-grained material will not work. Once a supply of the gravel pack is secured, the designer should obtain several representative samples of the pack for analysis. The gravel pack should be checked for grain-size distribution, particle makeup, and perhaps even a leach test for water quality.

The openings of the intake in all gravel-packed wells should be the 90 percent retained (10 percent passing) grain size of the selected gravel pack. This will allow adequate development without removing too much of the gravel-pack material itself (fig. 5-5). Selecting a larger opening will result in gravel-pack production during the development stage and perhaps a breaching of the gravel-pack filter by the fine aquifer material. Choosing a slot opening smaller than that recommended will not allow the well and gravel pack to be fully developed and cleaned.

An often misunderstood type of gravel pack that is used in the water-well industry is the stabilizing pack. A stabilizing pack is noth-

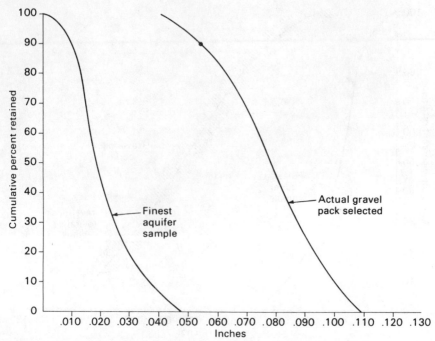

5-5. Sand analysis of finest aquifer sample and the actual gravel pack selected. Intake opening will be 0.054 inches.

ing more than a filler. The intake is designed as if the stabilizing pack were not present, but instead of allowing the formation to collapse around the intake during development, the void area between the intake and aquifer is filled with a coarse gravel in an effort to keep the formation from either collapsing too hard around the intake or to provide some coarse material around the intake in an effort to create a zone of high permeability. A stabilizing pack need not be designed in any particular way.

INTAKE LENGTH

Well efficiency is a relationship of the overall well-design factors and how the well fits into the aquifer. Given an amount of water that will be withdrawn from an aquifer, an intake that penetrates the full thickness of the aquifer will yield that water more easily than one that penetrates only 50 percent of the aquifer's thickness. More head is needed to force the same quantity of water through a smaller area as well as creating a longer path line for some ground water to move along before it enters the well.

One obvious factor in the design of an intake for a well is that many of the intake parameters are already given. The well's diameter is a function of the amount of water that the well is expected to produce; the intake's openings are designed based on a set of rules that already have been discussed. The length of the intake is the only thing left for the designer to specify. It is also the only factor that the intake's price is based on, leading many designers to cut costs by cutting the length of the intake based on antiquated entrance velocity factors. The success or failure of a well is directly proportional to the amount of aquifer that the well penetrates. After reviewing Darcy's law of ground-water flow, it should be apparent that the length of the intake is not the place to cut corners (fig. 5-6).

Two different geologic settings exist that formulate the rules for the proper length of an intake. The first is the *unconfined* aquifer and the second is the *confined* aquifer. Each situation has a different set of hydraulic rules that govern ground-water flow into a well. Unconfined flow is primarily governed by gravity drainage whereas confined aquifer flow is governed more by piezometric head within the aquifer, both before and during pumping.

5-6. The success of the well often depends on the length of the intake.
(Water Well Journal Publishing Company, Columbus, Ohio)

Unconfined Aquifers

In an unconfined aquifer, flow into the well is caused by a lowering of the water in the well, causing water to flow into the well by the force of gravity. Intakes that penetrate these aquifers need to be designed to take full advantage of the gravity flow into the well, while exposing enough aquifer material so that acceptable water production can occur. Using a screen length equal to one-third to one-half of the total unconfined aquifer thickness will take advantage of both situations. Typically, the well is designed with a length of one-third of the aquifer exposed. The designer can then estimate the well's production based on the permeability and transmissivity data obtained during test drilling or estimated from samples of the aquifer. If the production falls short of the expected yield, the designer then increases the length of the well's intake until the optimum length is found. In no case should the length exceed 50 percent of the saturated aquifer thickness. The saturated aquifer thickness should be calculated based on the well's depth (or depth to the bottom of the aquifer) under low water-table conditions. By following this rule, the ground-water flow pattern will be such that the maximum production versus intake cost can be obtained.

Pumping equipment can be placed either above or below the intake, never within the intake portion of the well itself. Placing a pump in the intake portion may cause irreversible damage to the intake and perhaps complete failure of the well. If the pumping water level in the well is of concern, a sump or well extension can be designed into the well structure that will house the pumping equipment below the intake. If this is the case, special precautions to cool the pump need to be taken. Manufacturers of pumping equipment will outline the best possible method for accomplishing this. Placement of the intake should be in the lowermost practical portion of the aquifer for maximum specific capacity.

Confined Aquifers

The lengths of intakes used in confined aquifers are not as easy to calculate as those constructed in unconfined aquifers. The well is penetrating an aquifer in which the ground water is under pressure, so the length of the intake is crucial to the success of the well. Ideally, the intake can penetrate 100 percent of the confined aquifer, taking full advantage of the natural pressures that are built up in the aquifer. Many times the economics of such a proposition requires a compromise.

Kozeny discovered a viable compromise between efficiency and intake length for confined aquifers. Figure 5-7 shows the results of the research that Kozeny performed (Kozeny 1933). The designer selects a curve

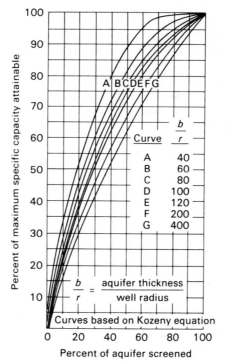

5-7. Relationship of partial penetration and attainable specific capacity for wells in homogeneous confined aquifers. (From Driscoll, 1986, © Johnson Division)

based on the relationship of the well's diameter with that of the aquifer's length. The curve is then followed until the percent of aquifer screen and the curve intersect. From that point, the designer reads the percent of the maximum specific capacity that he can expect from the well if he screens a certain percentage of the aquifer. The designer can adjust the length accordingly, noting that efficiency drops as the length of the intake decreases in relationship to the thickness of the aquifer.

The curves also demonstrate that there is a practical point at which the length of the intake versus the amount of extra specific capacity is no longer cost effective. In many cases, this point occurs around 80 to 90 percent of the screened portion of the total aquifer thickness. Summarizing Kozeny's curves, the optimum screen length for confined aquifers is as follows. When aquifer thickness is less than 25 feet, the screen should be 70 percent of the aquifer thickness for maximum specific capacity. When aquifer thickness is between 25 and 50 feet, the screen should be 75 percent of the aquifer thickness for maximum specific capacity. And when the aquifer thickness is 50 feet or more, the screen should be 80 percent of the aquifer thickness for maximum specific capacity. These lengths take into account the effect of partial penetration into a confined aquifer.

Placement of the intake should be in the most permeable zone of the aquifer unless the percentage screened is less than 75 percent. In that case, the intake should be dispersed into two or more sections, screening the most permeable zones first by separating the intakes with pieces of blank pipe. This practice will take full advantage of the natural pressures that drive ground water into the well.

6

Constructing Water Wells

An overview of well construction, from site preparation, to safety practices, to grouting, to site clean-up procedures, helps in the understanding of the time and economics involved in working toward a successful well-drilling operation. Construction of water wells is by no means an exact science. Technology and first-hand experience by the well driller are constantly changing the finer aspects of well drilling and well construction. New techniques are developed through field experimentation and through data that has been collected from formation samples, aquifer samples, and laboratory studies. So, while the face of well construction techniques remains basically the same, the underlying facets are undergoing many transformations.

The majority of well drillers are attuned to new developments. Hard and fast rules do not always exist in the well-drilling operation. Formations vary in position, consistency, and permeability as well as other characteristics. Most drilling axioms and guidelines do remain constant, but when adjustments become necessary they should be implemented to achieve a well of good quality at the most reasonable cost. Thus, planned well construction is often modified in the field. Just as with any type of construction, plans are made and blueprints followed, but the final assembly is more often than not subject to modifications and improvements.

PERMITS AND REGISTRATION

Before the well construction, the contractor must comply with state and local regulations. The drilling contractor usually has to be registered and licensed with the state and/or county. In some states a "start card" (stating location, type of well, owner's name, and so forth) must be filed before any drilling activity occurs. Upon completion of the well, a well-completion report may have to be filed. This report usually details all well information, including location, depth, yield, and strata encountered. As stated in the Environmental Protection Agency's *Manual of Water Well-Construction Practices*, article 10:

> The contractor shall give all notices and comply with all laws, ordinances and rules and regulations applicable to the work.... the con-

tractor shall pay all sales, consumer, use and other similar taxes required by the law of the place where the work is to be performed.

After complying with permit and registration requirements, the contractor begins to assess the drilling site. Difficult access and maneuverability of the rig can cause delays and added costs to the operation. Problems may arise with legal or physical access. Rights-of-way to the site may be necessary, and this responsibility lies with the owner. Likewise, any easements that are required for permanent structures or changes to existing facilities should be secured and paid for by the landowner. Also, an understanding should exist as to how much (if any) damage to the landscape (ruts, tree or shrub removal) will be permitted.

SITE PREPARATION AND SAFETY

The terrain and stability of the land should be considered by the contractor. Ideally, the well driller is looking for a relatively flat terrain that is firm under the rig's weight and has few obstacles. All of these conditions are not always present, so decisions are made as to the exact drilling site that presents the least amount of potential difficulties. And depending on the type of drilling, the well depth and subsurface formations, a choice can be made on the size and type of rig. For example, if a number of wells are to be drilled in rough terrain, a smaller, more portable drilling rig should be considered.

The stability of the ground not only affects actual drilling operations, but directly affects the rig set-up before any drilling begins. Proper and adequate blocking of the portable drill rig is a must; any time taken during this phase is well spent. Improper blocking is a major cause of drilling rig losses and an extremely hazardous condition that can lead to serious injury.

The blocking procedure begins when jacks and/or outriggers are extended from the rig to the ground, raising the rig partially or entirely off the ground. Blocking is then placed between the jack swivel and the ground to more effectively distribute the rig's weight. Proper blocking ensures that differential settling does not occur. Scrap lumber found on the job site should not be used; a set of permanent blocking devices should be kept on the rig at all times. Adequate blocking materials should measure at least 2 feet by 2 feet and be of sufficient strength.

Most types of drilling require at least some water and often this can be supplied from a water tank truck. However, for some drilling operations, a water supply at the drilling site may be necessary. For example, reverse-circulation and auger drilling sometimes require vast amounts of water. A portable pump and lengths of water hose might be needed to bring water in from a nearby river, or a bulldozer may be needed to excavate a holding tank for adequate amounts of water.

The quality and pH of the water supply should be tested. To avoid the contamination of aquifers (for example, in jet drilling) the water supply may require treatment with chlorine or other chemicals. The pH levels should be adjusted with soda ash. Other expenses may occur because poor-quality water cuts down on the efficacy of drilling muds. More additives may be needed to maintain optimum drilling rates.

Avoiding any contact with overhead power lines should be of primary concern. Serious injury or death from electrocution can occur in an instant. The most frequent cause of death in the water-well industry is electrocution caused by contact with overhead power lines while raising the rig's derrick structure. Every attempt should be made to choose a drilling site that is at least 20 feet away from any power lines or electrical structures. Often, drilling locations can be moved to a safe area, but in some cases, the positioning of a rig close to power lines is unavoidable and stringent safety precautions are imperative (fig. 6-1).

Even though power lines appear harmless, all electrical wires should be considered alive and potentially dangerous. Electric current always flows to ground when a suitable path is provided. When the derrick of a drilling rig touches an overhead wire, the entire drilling rig becomes electrified. The rig may appear safe, but its rubber tires insulate it from the ground; if a crew member standing on the ground touches the rig, a ground is then provided to the deadly current. As one helpful precaution, all control handles for operating the rig should be insulated with rubber grips or heavy wrappings of electrical tape. This may not prevent injury, but will lessen the severity of the accident. However, nothing takes the place of complete awareness by drilling personnel.

The contractor must be aware not only of overhead lines, but of underground power lines as well. If underground lines are hit, delays in drilling operations usually follow and the same possibility exists for serious injury or death. The local power company should be consulted. Usually a qualified representative will be sent out free of charge to advise the drilling contractor.

Electrical shock during the installation and testing of electrical water pumps is a hazard. This danger must be dealt with directly by the well driller or pump installer; careful procedures and common sense will eliminate most of the risks. The only electrical parts that should be worked on are those known to be disconnected from power. Before connecting power to the completed pump and control circuit, all circuits should be ground-checked.

Of course, safety practices at the drilling site go beyond the instances just mentioned. Safety precautions should pervade each state of the

108 · *DESIGN AND CONSTRUCTION OF WATER WELLS*

6-1. Every safety precaution should be taken when drilling near power lines. (Water Well Journal Publishing Company, Columbus, Ohio)

drilling operation, from rig set-up procedures to site clean-up details. The alternative to safety consciousness is the injury and layoff of valuable personnel and consequent delays in the drilling schedule. Well drilling is a potentially hazardous occupation, but proper steps can be taken to minimize the risks.

Drilling personnel should wear the appropriate safety equipment such as heavy steel-toed work boots, hard hats, and gloves (fig. 6-2). Because of the constant revolving of the drill-rig machinery and drilling equipment, loose clothing should be avoided. Personnel should maintain as safe a distance as possible while working near this moving equipment.

When starting drill-rig motors, operators should have the brakes set and the gears in neutral. When starting the air compressor motors,

Constructing Water Wells • **109**

6-2. Proper safety equipment should be worn at all times. (Water Well Journal Publishing Company, Columbus, Ohio)

the receiver tank valves should be open. Gloves should be worn while disconnecting the hot drill rod from the drill head. Machine power controls should be such that they cannot be accidentally activated.

Proper maintenance and inspection of all drilling equipment and rig machinery is essential. All gears and moving parts should be greased or lightly oiled to keep them working properly, to avoid rust, and to avoid breakage and possible injury. All drill rods, joints, cables, hoses, and so forth should be visually and physically inspected periodically. Screws, nuts, bolts, and any other connections should be checked for tightness on a regular basis.

The site should be secured from onlookers or vandals while operations are temporarily shut down. If the rig is left onsite overnight, possibly warning lights and/or a protective fence should be installed. All tools and valuable equipment should be secured to avoid damage or theft.

HOW WELLS ARE CONSTRUCTED

Before any well is drilled, certain plans are drawn up as to the size of casing to be used, type of screen, and probable depth to a water-bearing formation. This is done as a preparatory measure and to facilitate cost estimating. Planning the well lets the well driller order the appropriate equipment (such as well screens) ahead of time so there will be no delays upon completion of the actual well-drilling operation.

After the hole is drilled, the next stages make up what is known as the construction of the well. In a sense, the drilling of the borehole can be compared to laying the foundation of a house or building; a properly laid foundation ensures a good start to a sound and durable building structure, just as a properly drilled borehole sets the stage for a long-lasting and good-yield water well.

As the well-drilling operation progresses, plans and methods are generally altered or modified somewhat due to drilling conditions and the type of formations encountered. The well driller, the engineer, and any other involved persons must be flexible in this regard; little is accomplished when plans are rigidly followed in spite of indications that suggest the modification of such plans.

The four main parts to constructing a well include installation of (1) the permanent casing, (2) the well screen, (3) the gravel pack, and (4) the grouting. The development and disinfecting of the well are equally important aspects of well construction and will be addressed later in this chapter. Installing the casing and the well screen are basically the same procedure, sometimes being installed as one unit. Gravel packing comes next and grouting placement follows, rounding out this particular stage in well construction.

The permanent casing is what makes the well stand up over time, withstanding shifts in the earth and preventing cave-ins. The casing also helps to eliminate contaminants and the seepage of water from undesirable aquifers. The well screen or intake is the device that allows water to enter the well while preventing formation particles from doing the same (see chapter 5). The gravel pack assists in the development of the well, maximizing water yield while helping to prevent future problems such as sand pumping and clogging of the screen openings. Finally, the grouting procedure further seals the well from contaminants and helps to stabilize the casing and the surrounding formation. Collectively, these four parts are the structure of the water well.

Preparing for Installations

Before the installation of the casing/intake assembly, the well driller must first be equipped with the necessary information about downhole conditions. This information contributes to the ultimate success of the well. At this point, the borehole walls should be relatively stable and any cobbles or sand heaves should be bailed out. Any obstruction that might impede the setting of the casing and the proper placement of the well screen should be eliminated.

At this stage, all the data gained from analyzing the formation samples becomes very important. For this reason, the well driller must keep an accurate driller's log concerning the formations encountered and the depth to water-bearing formations. Inaccurate or incomplete data can result in the installation of inadequate materials, can delay completion of the well, and can eventually cause the need for drilling a new well. Many wells capable of producing water are abandoned because of incomplete drilling.

Recognition of Water-bearing Strata

There are no set standards for recognizing water-bearing strata. If little is known of the geology of the area, all possible water-bearing zones should be tested. However, sticky clays, tough plastic shales, thoroughly cemented sandstones and conglomerates, and massive unfractured limetones or other dense rocks seldom yield usable quantities of water.

As drilling progresses, an experienced well driller can make a rough estimate of the permeability of the formation by examining the drill cuttings as they reach the surface. With certain drilling methods, such as cable tool and air rotary, a well driller can easily detect the depth at which a water formation has been penetrated. However, with the conventional rotary method, the presence of drilling mud can make the detection of aquifers more difficult. The following criteria, based on changes in the character and circulation of the mud, are some of the most important means of recognizing water-producing zones:

1. As discussed in chapter 3, there is usually a slight loss of circulating mud while drilling caused by natural seepage into the surrounding formation. The continuous addition of small quantities of water is required to maintain the proper volume for circulation. However, when the loss becomes excessive, the depth at which the loss occurs should be noted carefully and formation samples taken frequently at that depth. A suitable screen can then be set at this level.
2. A noticeable increase in the volume of the returning circulation fluid invariably means that an artesian water-bearing formation has been penetrated, which has a higher hydrostatic pressure than the mud column in the hole. When this occurs, the circulating mud rapidly loses viscosity and weight and, in extreme cases, is displaced completely by water coming from the formation.

3. Complete loss of circulation usually means that a very porous formation has been entered, and the ground water in that strata is being pushed back by the drilling fluid invading the formation.

The proper testing of a suspected water-bearing zone can result in a desired yield, which in turn can eliminate the need for further drilling. If the well is being drilled in a hard, noncaving formation, the drilling mud should first be displaced by circulating clear water through the drill pipe to the full depth of the hole. Water then can be bailed out of the hole, thus lowering the fluid level and allowing the water in the suspected zone to come into the hole. This will wash out some of the mud sealing the formation at that point. The yield then can be estimated from the rate of bailing. If the well is being drilled in a soft caving formation, displacing the mud with water will result in slumping or caving and this method cannot be used.

Another procedure involves drilling a pilot hole (or small-diameter hole) until a suspected zone is penetrated. When such a zone has been reached, the hole is reamed out to a full size to within a few feet of the suspected zone. Then, a test string of casing is set, which will pass freely through the reamed portion but will not enter the pilot hole. The temporary casing is driven firmly into the formation that immediately surrounds the pilot hole, thus sealing off the bottom of the reamed portion of the hole. The annular space outside the casing is kept full of mud fluid to maintain the hydrostatic pressure on the borehole wall and prevent caving.

The drill pipe or wash pipe is then run through the casing to the bottom, and the mud in the suspected zone may be washed out with clear water. The fluid inside the casing can be bailed out to test the volume and quality of the water from the suspected zone. While bailing, the level of the mud fluid outside the casing should be watched carefully to make sure that the bottom shut-off is satisfactory and that all the fluid being bailed out is coming exclusively from the formation under test. If the mud fluid level outside the casing should drop, the casing will have to be reseated. After completing the test, the casing should be filled with drilling fluid before it is pulled from the hole. This fluid will prevent caving and permit drilling deeper after the test, if desired.

Accurate measurements to depth of water must be taken. Measurements to the static or pumping level are important and different methods can be used with accurate results, including the use of an air line, electric sounder, or the simpler wetted-tape method. Although only practical to depths of about 100 feet, the wetted-tape method can be used readily and with accuracy. In this method a lead weight is first attached to the end of the tape and the last 2 or 3 feet of the tape is wiped dry and coated with carpenter's chalk or keel. The tape is lowered into the well until a part of the chalked section is below the water. One of the foot marks is held exactly at the top of the casing and the measurement is recorded. After the tape is pulled up, the wetted line on the tape can be read to within a fraction of an inch, and this reading can be subtracted from the previously recorded foot mark made at the top of the well.

Coordinating Equipment

As a preparatory measure before installing the casing/intake assembly, all casing should be at the drilling site far enough in advance so as not to cause delays. The well screen, lead packers (if used), and all other necessary equipment should be at the site also. If PVC casing is to be used, care should be taken in storing and handling the sections, because PVC damages easily. To avoid heat damage, PVC pipe should not be stored in direct sun. If steel casing is used, all welding tools and equipment should be ready.

Some drillers install centering guides on the side of the casing (fig. 6-3). These guides can make it easier to install the pipe. The guides tend

6-3. Centering guides used during screen installation. (Water Well Journal Publishing Company, Columbus, Ohio)

to center the string of casing in the hole, ensuring an evenly spaced, annular void, which will result in uniform grouting seal. Some drillers say that the borehole is too irregular and that centering guides are of questionable merit.

Hoisting the casing and lowering it into the hole requires some preparation. The casing should be stacked fairly close to the drilling site in a manner that will facilitate the removal and lifting of each length of casing from the stack. To the hoisting cable is attached a hoisting plug (used for lifting drill pipe), which by means of a suitable adapter can be used to lift sections of casing and lower them into the hole. The plug has a bumper wheel that enables the operator to make or break joints quickly and easily. The hoisting plug should be kept clean and the threads and bearings should be oiled as necessary.

The well driller may employ a safety clevis, whereby two 1-inch holes are torch cut into the upper end of the casing, and a safety clevis is attached to each hole. After the casing has been lifted and lowered or driven into the well, the clevises are removed, and the two holes are welded shut before the casing is set completely into the borehole.

Intake Installation Methods

Procedures for installing the well screen can vary according to the drilling method used and the design of the well itself. Every method requires taking accurate and complete measurements of pipe length, screen length, cable length, and depth of hole (fig. 6-4). Again, the particular method originally planned for a well may have to be changed according to problems encountered in the drilling operation. The commonly used methods of screen installations are discussed here.

Pull-Back Method

The pull-back method is the simplest and best way to install a screen regardless of the drilling method, but it is particularly applicable to cable-tool drilling. In this operation, the well casing is sunk to the hole bottom and any sands or other particles are cleaned out with a bailer. If problems arise from the sands heaving inside the casing, the situation usually can be remedied by adding water to keep the water level in the pipe above the static water level in the water-bearing formation.

In the bottom of the screen, a bail plug or heavy plate is installed to which a screen hook is attached, allowing the screen to be lowered into the well via a sand line. A lead-packer fitting is screwed to the screen's top and the screen is then lowered inside the well casing. After setting the screen on the bottom, the casing is pulled back far enough to expose the screen in a water-bearing formation, but only to a position where the lead packer is still about 6 inches to 1 foot up inside the

6-4. Accurate measurements help to ensure the efficiency of the well. (Water Well Journal Publishing Company, Columbus, Ohio)

casing. A casing ring and slips, together with two hydraulic jacks, are usually needed for pulling the pipe. If the screen moves upward as the pipe is pulled, the drill bit or other tool can be lowered into the well to hold the screen in place.

The lead packer is then expanded inside the wall of pipe using a swage block. The block is lowered into the well and seated in the lead packer. The sliding bar that runs through the center of the swage block is raised and dropped several times; these light blows will expand the packer. The bar should not be raised far enough to lift the block out of the packer while swaging. Only light taps are needed to deform the lead ring and make a good seal.

Telescoping well screens were originally developed to facilitate the setting of screens in wells drilled by the cable-tool method. As the name suggests, the top portion of the screen is made just the right diameter to telescope through standard pipe of the corresponding size, that is, a 6-inch telescope-size screen is just the right diameter to pass through 6-inch standard pipe. The screen diameter is designed for just enough clearance to prevent sticking.

Self-Sealing Packer

Before 1964 the standard method of sealing a telescoped well screen inside the casing was to expand the attached lead packer. Then a self-sealing packer was introduced by the Johnson Division that consists of a flexible neoprene sealing ring (Johnson Division 1970). The ring is securely mounted on a heavy metal fitting that is screwed or welded to the top of the screen or to an extension pipe above the screen. The packer is called self-sealing because, unlike a lead packer, no swedging or expanding operation is necessary. The design is such that the packer seals the annular space between two cylinders. It resists being displaced to one side or the other even under considerable lateral pressure.

A well screen, fitted with a self-sealing packer, is installed through the casing or, in some cases, through a liner by simply lowering the screen assembly with the packer attached. If the weight of the screen assembly is not sufficient to overcome the friction of the packer inside the casing, the added weight of a bailer or other tool may be used to help push the screen assembly to the bottom of the well. Limitations to the use of self-sealing packers are:

1. They are not intended for use when the screen installation requires that water recirculate on the outside of the screen.
2. They are not designed for sealing a liner or casing in an open borehole in a consolidated formation.
3. Casings in poor condition or inadequate in size are likely to prevent proper setting/sealing of the packer.

Another method of installation is to drill an open hole ahead of the casing to receive the screen. The casing is driven into the water-bearing sand formation to a depth that is a little below the desired position for the top of the screen. Drilling mud is then mixed and the casing is filled with the fluid. An open hole is drilled in the water-bearing sand formation beyond the end of the casing deep enough to accommodate the length of well screen to be exposed. The drilling mud should be heavy enough to prevent the open hole from caving during later development.

The screen is lowered into the extended hole, making sure that the packer is still inside the wall of the casing. The end fittings for the screen used in this method are the closed bail plug and the lead packer top. If the hole is too deep, gravel can be dropped in to fill the hole to the proper height. When the screen is in proper position, the packer is expanded and development of the well proceeds.

Bail-Down Method

Special fittings for the screen are required for this method. The well screen is fitted with a bail-down shoe or with an open sleeve at its lower end, and is then lowered down through the casing. With special connection fittings, the bail-down shoe and the screen may be suspended from a string of pipe called the bailing pipe.

The entire assembly is lowered into the casing until it reaches the targeted formation. The bailer and drilling tools are operated through the screen or drilling pipe, whichever the case may be. The bailer is then employed to remove sand from below the screen, allowing the screen to settle into position. As the screen is bailed down, the operation should progress continuously to prevent the sand formation from packing tightly around the screen and impeding its downward progress.

If using the bailing pipe (and after the screen has reached the desired depth), a lead plug is dropped through the bailing pipe to seal the opening in the nipple above the bail-down shoe. The entire string of bailing pipe is then unscrewed to the right and disconnected. After the lead packer is expanded, the well is ready for development.

Setting Screens in Rotary Drilled Wells

The pull-back method as previously described can be used in a similar fashion for wells drilled by the conventional rotary method. The primary difference in this type of installation is that the downhole casing must be suspended in place during screen installation and development by using pipe clamps or a spider and slips, because there is little or no friction around the outside of the casing to hold it in place.

Wash-Down Method

This method is similar in principle to the bail-down method, except that instead of bailing procedures, water (not drilling fluid) is circulated through a wash pipe to clean out the material underneath the screen, thus allowing it to settle into the proper position. After stopping the pump, time should be allowed for the water-bearing sands to close in around the screen. When the formation has enough friction to hold the screen in place, the entire string of wash pipe is turned to the right and disconnected. The lead packer is expanded and development of the well can begin. This method of installation works best where the water-bearing formation is composed of fine to coarse sands with little or no gravel.

The screen and casing can be made up in a single string assembly and set in the drilled hole, omitting the wash-down operation. The screen fittings required are a pipe thread fitting at the top and a closed bottom. When the entire string has been run into the hole, pipe clamps should be used around the casing at the surface to carry all or most of the weight until the formation closes in around the screen.

The single-string method of installation is suitable for shallow wells up to 50 feet. For deeper wells, the telescope type of screen installation is much better because it permits the grouting or cementing of the casing in the hole before the screen is installed. Proper grouting of the casing is impossible with single-string installation, and an outer casing of larger size must be used if grouting is required. The telescope method also permits the screen to be removed and replaced when necessary. Furthermore, using this method eliminates the bad construction practice of setting a long string of casing on top of a long well screen. When the screen hits bottom, it becomes a loaded column that is easily buckled because of its slender structure.

Plumbness and Alignment

A hole drilled into the earth to any substantial depth will not be perfectly straight or perfectly plumb. Some deviation must be allowed in the construction of water wells. However, the drilling contractor should exercise great care and patience to achieve a borehole that is within practical limits.

There are several methods suitable for checking the plumbness and any deviations. Many modern water-well specifications require that plumbness be checked with a specially designed plumb bob, and that straightness of the hole be tested with a 40-foot cylindrical dummy, slightly smaller than the inside of the well casing. Of the two factors, straightness of the hole is considered to be the more important, because a pump can be installed without difficulty in a well that is reasonably straight but out of plumb. However, the drilling contractor must keep in mind that too much deviation from the vertical may affect the operation and life of some pumps, so plumbness should be controlled within reasonable limits.

The conditions that cause wells to be out of plumb are:

1. Character of the subsurface material penetrated during drilling
2. Trueness of the pipe used as a well casing
3. Pull-down force on the drill pipe during rotary drilling

While drilling, gravity assists in making the drill bit cut a vertical hole, but varying hardness of the material being penetrated deflects the bit from a truly vertical course. The edge of a boulder in a glacial drift, for example, will force either a cable tool or rotary bit to one side. In cable-tool drilling, the boulder may deflect the well casing as it is driven and cause the hole to slant as the well is deepened. Variations in the hardness of consolidated rock formations also will start a deflection, resulting in a continued drift of the hole from the desired alignment.

Gravel-Pack Installation

Artificial gravel-packing of wells is a common practice that results in higher yields from the aquifer. The water demands of the rice industry in the early 1900s gave rise to gravel-packing water wells; at that time irrigation wells were gravel-packed. Popularity of the practice has increased with the advent of reverse-rotary drilling. The gravel packing of drilled wells using an outer casing, and wells drilled by either reverse-circulation or the standard rotary method, have met with equal success in the petroleum and ground-water industries.

With artificially packed wells, a zone of natural formation immediately surrounding the well screen is removed and replaced by artificially graded coarser materials. Naturally developed wells retain the natural formation except that the finer materials surrounding the screen are removed during well development. In either case, a more permeable zone is created, resulting in the increase of the effective diameter of the well.

Two types of gravel packing are in general use: the uniform grain pack and the graded grain-size pack. The former has been widely accepted since the 1960s, especially when manufactured screens are used, because the size of the openings can be controlled.

Probably the most common cause of sand pumping wells is the use of a gravel pack that is too coarse for at least part of the formation. Sandwiched between coarser sand and gravel particles, a relatively thin interval of fine sand may continue to sift through the pack indefinitely. This problem is caused by (1) poor sampling, and (2) lack of care in selecting the gravel-pack size. The gravel (or sand) pack should be selected as a ratio of the finest part of the formation to be screened.

Problems may exist with graded grain-size packs in that the pores of the gravel-formation interface may become partially blocked by the surrounding formation material; the result is reduced permeability. By using a well-sorted uniform gravel pack, the fines do not plug up the interface, but rather are drawn into the well during development. Formation permeability is increased while retaining the highly permeable nature of the pack.

The four most common reasons for gravel packing are:

1. To increase the specific capacity of the well
2. To minimize sand flow through the screen in fine formations
3. To aid in the construction of the well

4. To minimize the rate of incrustation by using a larger screen-slot opening where the formation is relatively thin but very permeable and the chemical characteristics of the ground water suggest a potential for significant incrustation

During installations of graded gravel packs, the separation and segregation of the fine and coarse particles may cause sand pumping problems later. This separation of gravel materials can be minimized significantly by choosing the proper method for placement of the gravel pack. The segregation of materials in uniform grain-size packs is less likely to be a serious problem because the size of the particles varies only slightly. However, graded packs have the greater potential for separation of their wide range of particle sizes.

To illustrate separation difficulties, lab tests show that a round particle of a given size will fall through water four times faster than a particle half as large. Applying this axiom to the placement of a uniformly graded pack, single grains of 1/8-inch size will fall through the water and reach the well bottom four times faster than smaller grains of 1/16-inch size.

Therefore, with graded packs and uniform gravel packs, the material should be dropped into the annular space in lumps or slugs to minimize separation of the different size grains. A 2-inch or larger tremie pipe run to the hole bottom will help the well driller to accomplish this objective. Gravel is poured or shoveled into a hopper that is connected to the top end of the pipe, and a supply of water is usually poured down the pipe with the gravel to prevent any bridging of the particles downhole.

Gravel may be pumped downhole through the tremie pipe, which usually eliminates the need for water. This same procedure can be applied when installing gravel packs to deep wells where the tremie system is not practical. An advantage to this deep well pumping of gravel is that the proper amount of gravel can be placed in the annular space and overfilling above the placement of the setting tool does not occur. With the tremie system, the pipe is carefully pulled back up to the surface as the annular space gradually becomes filled with gravel. The well driller can probe the borehole with the pipe itself to determine how much of the hole has been filled with gravel. The tremie system of gravel packing is practical for shallow and medium-depth wells.

Whenever the tremie pipe method is not used and gravel is simply poured into the annular space at ground level, bridging is likely to occur, leaving a void below that point. To avoid this problem and to minimize particle separation, reverse-circulation methods are applied with the drilling fluid constantly circulating as the gravel is placed in the well. In this method, gravel is carried by the fluid down into the annular space outside of the casing. The gravel is directed to the well bottom around the screen as the fluid is drawn through the screen openings and up through the casing to the suction pump near the top of the well. During this time the borehole should be kept full of drilling fluid as recirculation operations are underway.

If the velocity of the descending stream in the annular space is about the same as the velocity at which a particle of gravel falls in a fluid that is not moving, no separation of sizes can occur. In a light drilling mud, the velocity of the fall of the particle will be less than one-half foot per second. The reverse-circulation of the drilling mud may be started at a relatively slow rate. The pumping rate will often depend on the flow of mud through the well screen openings, and water should be added to thin the mud so that circulation can be increased if necessary. As soon as the desired circulation is attained, gravel may be put in the fluid stream. The two limiting factors of the pumping rate will be the head loss of the fluid flow within the well itself and the suction capability of the pump.

Thinning of the mud with water does not increase the risk of formation caving since the descending fluid stream in the annular space tends to exert more borehole support even though the weight per gallon has been reduced. Once the introduction of the granular material is started, the weight of the material builds up the effective weight per gallon of the fluid; therefore, caving is seldom a problem.

After gravel-packing installation, some type of seal is usually required over the pack. This seal prevents the gravel pack from shifting and moving upward when the well is pumped. Cement grout in small sacks or lead shot can be placed above the gravel. Sometimes a lead-slip packer is placed over the top end of the extension pipe. This packer is then expanded by using a regular swedge block to complete the seal.

The development of gravel-pack wells has some difficulties. Due to the nature of the pack and the adjoining compacted natural formation, water forced through the well screen tends to slosh up and down throughout the gravel pack instead of shooting out horizontally into the formation material. For this reason, gravel packs should be designed as thin as possible so the sloshing effect will be kept to a minimum. The presence of a gravel pack also hampers the effectiveness of chemicals used to break down the drilling fluid that has penetrated into the aquifer during the drilling process. The larger borehole dilutes the chemical mixture, adding to the cost of development.

Grouting and Sealing the Well

Increased public awareness, extensive research, and stricter drilling codes have all brought about a widespread acceptance of grouting procedures for water wells. Since most contamination enters a well at the ground surface or near the surface, the grouting of water wells is a logical necessity if the quality of ground water is to be protected (fig. 6-5). Grouting and sealing the well, installing a proper seal on the top of the well, and disinfecting the well are all important procedures that will ensure a safe well.

Cementing or grouting a well involves filling the annular space between the outside of the casing and the inside of the borehole wall with a cement or bentonite grout mixture. When well construction has included both an inner and outer casing, grouting may be required in the annular space between the two casings as well as in the space outside the outer casing.

Planning the diameter of the hole is important because when improperly centered casing contacts the borehole wall, the slurry tends to channel, causing tight areas and dead spots. The actual size of the

6-5. The protection of ground-water quality is essential. (Water Well Journal Publishing Company, Columbus, Ohio)

grout space required usually depends on the method of grouting to be used, and this decision should be made prior to drilling. The area of the annular space around the casing frequently influences the success of the work and the completeness of the seal.

Grouting and sealing the casing in water wells is done for the following reasons:

1. To prevent seepage of polluted surface water down into the well along the outside of the casing
2. To seal out water of unsuitable chemical quality in the strata above the desirable water-bearing formation
3. To stabilize and secure the casing in a drilled hole that is larger than the pipe used
4. To form a protective sheath around the pipe, increasing its life by protecting it against exterior corrosion

Various forms of grouting can be used to serve these purposes. A cement slurry—a mixture of portland cement and water—is commonly used because it is inexpensive and its availability is not dependent on other service industries. Puddled clay may be used, but only if used at a sufficient depth where drying and subsequent shrinkage of the mud does not transpire. In addition, this clay mixture should be placed at a depth where water movement will not wash away the clay particles, rendering the grouting seal less than totally effective.

As another grouting additive, bentonite is a colloidal clay that requires a large volume of water. About 6.5 gallons of water are mixed with a standard sack of cement. Depending on the percentage of bentonite (from 1 to 8 percent) added to the cement, a slurry lighter than the neat cement will result with an increased yield. The addition of bentonite will reduce the compressive strength of the material. The bentonite and water should be mixed before the cement is added.

A significant advantage of a bentonite mixture is that less shrinkage (if any) occurs while the mixture is curing (as compared to a strictly cement mixture). Of course, even limited degrees of shrinkage can cause a thin but continuous void all the way down the annular space, creating a channel for contaminants to trickle into the well. Bentonite helps hold cement particles in suspension and improves the fluidity of the mixture.

Other additives to cement slurries include pozzolans, perlite, diatomaceous earth, and gilsonite. These additives are used to meet specific well conditions and perform various functions such as lowering the permeability, reducing the density, increasing the yield, and increasing the weight of the slurry. Often, economic factors will play a role in the choice and proportion of grouting mixtures, but the fulfillment of certain specifications may override any cost considerations.

Additives are mixed with neat cement to increase yield per sack, to reduce cementing costs, and to alter slurry properties for special well conditions. Additives are used as the following:

1. Extenders to provide a greater yield or slurry volume for each sack of cement
2. Weighting materials to increase slurry density and overcome high formation pressures
3. Accelerators, such as calcium chloride, to reduce waiting time in shallow-well completion, particularly in cold climates
4. Retarders, commonly used in wells 8,000 feet and deeper, to increase thickening time
5. Low-water-loss additives important to the "squeeze" cementing technique where control of water loss is critical
6. Lost-circulation materials to reduce loss of cementing fluids to a very porous or permeable formation, fractured zones, or weak formations hydraulically fractured by the pressure of mud and cement in the annulus
7. Special additives to change cement-flow behavior

Mixing Grout

For cement grouts, each 94-pound sack of cement requires the addition of 5 to 6 gallons of water, which should be mixed thoroughly to produce a suitable grout. More specifically, laboratory tests indicate that 5.4 gallons of water are necessary to hydrolyze one U.S. sack of cement. Mixtures of more than six gallons per sack are not acceptable for water-well cementing; shrinkage increases with water content. Water is removed from grouting mixtures by filtration through fine sand or other permeable formation materials.

An important advantage of the proper water-cement ratio is that more effective bridging of the cement particles occurs over the pores in permeable formations. With the correct mixture ratio, the cement is less likely to penetrate excessively into the adjacent formation material. Less of the mixture is wasted and costs are minimized.

For exceptional borehole conditions, the addition of sand or other bulk material may be necessary to bridge any large openings and prevent excessive fluid loss. These coarser materials generally make the handling and placement of the grout more difficult, but are a good choice to control costs and unnecessary fluid losses.

Any sands or other additives should be on hand and ready to mix as it is very important that grout placement is accomplished in one continuous operation. To ensure an effective grouting seal, the placement of the grout should not be interrupted. If, for example, the initial pour begins to set up before any additional grout is added, the grouting envelope will probably not be homogeneous throughout.

When fluid losses do occur, or when grouting is done for a deep well, larger quantities of grouting will be necessary and the equipment and materials for mixing should be dependable and adequate for the job at hand. Preparations must be made to ensure that the mixing apparatus does a thorough job at a rate fast enough to maintain continuous placement of the grout from start to finish.

The exact volume of grout that a well requires is often difficult to determine. Large cavities and other irregularities such as rock fractures downhole will render any standard calculations incorrect. The well driller must be prepared to meet a potential need for grout by initially mixing a slightly more-than-adequate amount, or by having the necessary materials and time available to mix any additional amounts before setting occurs.

Water used for mixing the grout should be free from oil or other organic materials. Dissolved mineral content should be less than 2,000 parts per million. Water having a high sulfate content is particularly undesirable for grouting applications.

Grouting Placement

The placement method for grouting can determine the effectiveness of the seal. For best results, the grout should be introduced via grout pipe at the bottom of the interval to be grouted. This procedure not only minimizes possible bridging (which can occur when grouting is introduced from above), but also minimizes contamination or dilution of the slurry.

Initially, the grout pipe should extend to the bottom of the annular space, and the delivery end should remain submerged in the slurry during the entire placement of the grout. The pipe may remain stationary or be gradually pulled upward as the space is filled. Should grouting operations be interrupted, the grout pipe should be raised above the grout level; only after all water and air have been displaced from within the grout pipe should it again be lowered into the slurry. In this method, suitable pumps, air pressure, or water pressure is used to force grout down into the space to be filled. Placement by gravity is an alternative method and may be satisfactory and more practical in some cases.

Gravity placement in a shallow borehole, for example, is accomplished by introducing the slurry to the hole bottom. After the borehole has been filled partially with grout, the casing—equipped with centering guides and with a tight, drillable plug to seal the bottom opening—is lowered into the slurry. As the casing is lowered, the grout is forced upward around the casing, filling the annular space. When the grouting has set, the plug is then easily drilled out from the casing and drilling can continue below the grouted interval.

A pipe of 2 to 4 inches in diameter should be used to conduct the proper amount of cement slurry to the bottom of an oversized hole. If the hole has been drilled by the conventional rotary method and still has drilling fluid in the hole, this mud will rise upward during grout placement because the cement slurry is heavier by density and weight. The volume of grout placed must be adequate to fill the annular space around the plugged, permanent casing as it is sunk to the proper depth. When the cement has set and sufficiently hardened, the bottom plug is drilled out. Drilling is then continued below the grouted section. A 72-hour setting period is normal for most portland cement slurries, but the waiting period may be reduced to 48 hours by using high early-strength cement.

Grout Pipe outside Casing

Grout may be placed through a string of small piping that is run down the outside of the casing, provided the annular space is of sufficient size to accommodate the grout pipe. First of all, the casing—with attached centering guides—is lowered into the hole. Either the lower end of the casing is sealed with a drillable plug, or open casing is driven into clay or other adequate formation material. Then the casing may be filled with water to prevent it from rising in the slurry, or the casing can be held down by the weight of the rig.

The grout pipe—generally 3/4-inch or 1-inch pipe is used—must be large enough in diameter to allow the required volume of grout to be placed in the time available. The oversized hole should be 4 to 6 inches larger in diameter than the casing in order to provide sufficient net space to accommodate the grout pipe.

Grout is placed by gravity flow only when it is certain that the entire operation can be completed quickly. Pumping is preferred because it facilitates the rapid introduction of the required volume of grout. A pump capable of delivering a pressure equal to the hydrostatic pressure of the grout plus the fluid friction in the grout pipe and, to a lesser degree, in the annular space, is usually required.

Grout Pipe inside Casing

When the use of small pipe outside the casing is impractical, grouting is accomplished by means of a grout conductor pipe installed within the casing. This procedure is referred to in the oil industry as the tubing method of cementing. A suitable packer connection—a cementing or float shoe—is used at the bottom of the casing to regulate the grout flow and to prevent reverse flow of the grout both during the procedure and after removing the grout pipe.

During the procedure the casing is suspended slightly off the hole bottom and should be weighted as before. While the grout is being introduced, a ball-type check valve prevents the reverse flow of grout into the casing. Upon completion of the cementing procedures and an adequate setting time, the packer material is drilled out of the casing. A variation of this, called the inside casing method, is done without a cementing plug or float shoe at the casing bottom. A fluid-tight stuffing box in a heavy cap that closes off the top of the casing is used.

Casing Method

In the casing method of grouting, the slurry is forced down the casing and into the annular space. Widely practiced in the oil industry, this method employs either one or two spacer plugs; one plug separates the cement slurry from the fluid in the casing and the other plug separates the slurry from water pumped in above the plug. After pumping water or mud through the casing to circulate fluid in the annular space and clear any obstructions from the hole, the first plug is inserted into the casing, which is then capped. A measured volume of grout is pumped into the casing. Then the cap is removed and the second plug is inserted, resting on top of the column of grout.

A measured volume of water is pumped into the casing until the second plug is pushed to the bottom of the casing, expelling the cement slurry (and the first plug) from the casing and into the annular space outside the casing. The water-filled casing is held under pressure to prevent backflow until the slurry has set. Spacer plugs are necessarily made of materials that can later be drilled out. Wood and rubber fiber-type plugs tend to elude the drill bit's teeth, so a plaster-type plug or equivalent is more suitable.

Modifications of the casing method, using either just the lower plug or just the upper plug, are based on the same principle of forcing the slurry into the annular space. Both are effective procedures for obtaining a proper grouting seal.

Finishing the Well

The final but crucial stages of well construction involve well development and the disinfecting of the well. Both procedures are essential, and if not done correctly or if not done at all they can undermine all the diligence that a well driller has exercised to this point. Development and disinfecting ensure the best possible water quality delivered at optimum yields.

Development

Developing a well simply involves those steps necessary to provide water in the aquifer with the most direct route to the well. Ideally, in unconsolidated formations a concentration of the coarsest sands and gravels is situated directly next to the well screen, with the degree of fineness gradually becoming more prevalent outward from the well. Depending on the surrounding formation, this gradation of particles continues for a few inches or for several feet. Water thus moves more freely as it enters the vicinity of the well. In consolidated formations, wells are developed by assuring free flow from fractures or, in some cases, by increasing the fractures through artificial means. Development work is most important in the completion of a screened well because the uniformity of the grading of the sand or gravel around the screen is improved by removing the finer particles. These fines are targeted for removal because they impede the flow of water into the well.

The chosen screen openings are large enough to allow the desired proportion of finer materials in the water-bearing formation to pass through the screen. The development work can exhaust the finer material, while the coarser materials remain stationary outside the well screen. The result is sometimes called natural gravel-packing of the well because the remaining coarser, more permeable material is actually part of the natural formation. In addition to improving the yield, proper development of a sand or gravel formation stabilizes the formation in the vicinity of the screen so the well should be relatively sand-free after completion.

There are two methods of development for screened wells. The natural development of a well involves the generation of a permeable zone around the screen solely by means of the development process. In this method, fine particles are drawn from the immediate area into the well, through the screen openings, after which the particles are removed by bailing or pumping. The remaining naturally developed zone of uniformly graded sand or gravel permits water to move toward the well screen with negligible head loss. The other method, called artificial gravel-packing of the well, involves the packing or placement of an envelope of artificially graded materials into a predetermined annular space between the well screen and the formation. Although some well drillers prefer to omit the additional development process in artificially gravel-packed wells, the well should be developed in order to maximize yields from the aquifer.

The fundamental purpose in each development operation is to induce alternate reversals of flow through the screen openings that will rearrange the formation particles, thus breaking down any bridging of groups of particles. During development the direction of flow is alternately reversed by surging the well; the outflow action of the surge cycle breaks down bridging that occurs during water intake, and the inflow action then moves the fine material toward the screen and into the well.

Regardless of the type of well or method of development, the first objective of the development process is to remedy any adverse alterations to the water-bearing formation that have occurred during the drilling operation. Every method of drilling plugs off a certain percentage of formation pores, thus reducing the conductivity of the aquifer to some degree. For example, when casing is driven during cable-tool drilling, the vibrations cause formation sands to settle and compact somewhat around the casing itelf. In reverse-circulation drilling, some clay and silt are picked up from the formation during circulation, but then may be redeposited as water loss occurs through other areas of the formation. The filtered-out clay and silt must be removed from water-bearing strata by means of development. This same clogging of pores occurs to a greater extent during conventional rotary drilling where the drilling fluid forms the filter cake as was discussed in chapter 3.

Development procedures require a good deal of time and patience on the part of the well driller. Repeated surgings bring progressively less and less amounts of sands into the well until an insignificant number of grains can be attracted through the screen openings. At this point the well is fully developed and should remain trouble-free for many years. Development methods are discussed in detail in chapter 8.

Disinfecting

Thorough disinfecting of the well is the last essential step in well construction and is necessary to ensure that there is no bacteria present. The disinfecting process leaves the water safe for drinking purposes and also prevents the aquifer from becoming contaminated. Unfortunately, the disinfecting of wells is often neglected by well drillers.

During the well-drilling operation, the material and tools used become contaminated with dirt and certain types of bacteria. Most of the germs that are picked up and introduced into the well are primarily harmless varieties that do not produce illness. But entry of any type of bacteria into the well can occur through contaminated drilling fluids via equipment or through surface drainage. For this reason, all newly constructed wells or any existing wells subjected to repair should be disinfected before being put into service.

When a certain germ, known as the coliform bacteria, is found to be present in the water, its presence indicates that the water has been polluted by either human or animal wastes. Coliform bacteria is used as an indicator because other disease-producing organisms are extremely difficult to detect. If coliform is present, then the water may contain other disease-producing organisms that normally live in the intestinal tracts of humans and warm-blooded animals. Water from a well is considered safe for human consumption only if tests prove no coliform bacteria is present.

Periodic disinfecting of a well during the drilling operation is recommended as a preventive measure. Each day the drilling progresses, the disinfectant can be added to the water in the well to disinfect the casing and the drilling tools. Theoretically, any water introduced into the well as a drilling fluid or for well stimulation should be of drinking water quality. Gravel-pack material should be disinfected before placement in the annular space.

A strong chlorine solution is the simplest and most effective agent for disinfecting or sterilizing the well, pump, and, equally important, the storage tank and piping system. Highly chlorinated water may be prepared by dissolving calcium hypochlorite, sodium hypochlorite, or gaseous chlorine in water.

Before disinfecting, the well should be cleaned as thoroughly as possible of foreign substances such as soil, grease, oil, joint dope, and so forth. These substances may harbor bacteria, especially the oil-based materials. A strong solution of chlorine should be introduced into the well in such a way as to ensure that all well surfaces above the static level will be completely flushed with the solution. The solution in the well should be agitated to uniformly distribute the chlorine throughout the well.

The most convenient method for preparing the chlorine solution is by dissolving one heaping tablespoon of calcium hypochlorite or sodium hypochlorite in a relatively small quantity of water. This solution should be thoroughly mixed to eliminate any lumps and then poured into one quart of water, which is left to stand for a short time.

One quart of this solution is enough to disinfect 1,000 gallons of water, so the amount of water in the well is first estimated and then the proper amount is prepared. Overchlorination of the well does not present any problems, so adding an excessive amount of the chlorine solution to the well is preferable to adding too little.

After thoroughly agitating the solution in the well by using a surging action, it should stand for several hours, preferably overnight. The well is then flushed to remove all of the disinfecting agent. When disinfecting an entire water system, the chlorine solution should be pumped through the piping with all faucets and valves in the open position to ensure a thorough and complete disinfecting process.

FINISHING THE SITE

Before the well driller drives away from the drilling site, certain steps should be taken to leave the area looking as much as possible as it did before drilling operations began. In most cases, the final preparations and cleanup duties should be the responsibility of the well driller. Thorough cleanup keeps the owner satisfied, restores the environment, and helps to instill a sense of pride among the crew members.

The site should have been kept relatively free from debris, loose tools, and other materials as the drilling progressed. Maintaining a clean, orderly work area contributes to the safety of the crew, the speed of the operation, and makes the final cleanup much easier. Before leaving the site, the area should be checked for trash, discarded hand tools, welding rods, sample bags, and so forth.

What to do with the cuttings from the well is an important consideration because the drilling of deep wells or large-diameter wells will result in significant amounts of cutting material remaining on the surface. If the terrain is rugged, cuttings may be bulldozed over an embankment if the owner permits, or buried in a large pit and then graded over with dirt. Some contractors shovel cuttings into the annular space outside the casing to a point of elevation no higher than the minimum grouting depth. In this way, the cuttings act as a filler material that saves on the amount of grouting used down the annular space. The landowner may request that the cuttings remain on the property, possibly for use as a driveway base or other similar project. Depending on the drilling contract, the owner may request that damaged lawns, shrubs, fences, or other structures be restored to their original state. All excavated pits or trenches should be filled in.

Before leaving the site, the driller must make sure that the top of the well has been sealed securely (fig. 6-6). Considerable time may pass after the completion of the well until the owner gets it completely

6-6. A securely sealed well further protects the ground water from surface contaminants. (Water Well Journal Publishing Company, Columbus, Ohio)

equipped and operating, so the installation of a secure cap assures the driller that the well remains in good condition. Capping of the well also prevents vandals from dropping any contaminants down into the well hole. A lockable cap is recommended. In areas prone to flooding, the cap must be watertight.

In public areas, the well should be protected by a fence or other barricade. In fields or tall grass, the well should be marked clearly with a tall stake or other device. It is important that attention be drawn to the presence of a well to protect it from damage. The casing in any well should extend at least 1 foot above the surrounding ground surface.

As another way to minimize surface contamination of the well, the immediate area around the well pipe should be graded with soil, creating a sloping effect outward from the well opening. This procedure discourages any seepage of drainage and polluted water down through the formation material immediately surrounding the well. This is particularly important when the well is drawing from shallow aquifers.

In some cases, a cemented surface apron may be installed around the well as a further preventive measure against contamination. However, improper surface-apron construction may cause serious problems with casing failures. Frost-heave pressures can transmit thousands of pounds of force against the casing. Radial expansion joints in apron construction are recommended to minimize these potential compressive forces.

If a pump house or service house is necessary to the well operation, the structure should be located 10 to 15 feet away from the well itself. This allows convenient workover of the well without endangering pressure tanks and electric plumbing configurations located in the pump house. It also allows room for a drilling rig to pull the pump if the need should arise.

7

Testing for Yield and Quantity

A pumping test can serve two different but related objectives. One objective is to describe the hydraulic characteristics of the aquifer. With this objective in mind, the test is sometimes referred to as an aquifer test. The other objective of a pump test is to provide information about the performance of the pump and the yield of the well. Such a test is often called a well test because the pump or well (rather than the aquifer) is being tested. Describing the hydraulic characteristics of an aquifer is the main concern here.

The principle behind an aquifer test is rather straightforward. Water from a well is pumped out at a specific rate for a certain period of time. The effects on the water table of this pumping are measured in the pumped well and in surrounding observation wells or piezometers. The hydraulic characteristics of the aquifer are found by substituting the drawdown data figures, the distance between the pumped well and the observation wells, and the well discharge numbers into the appropriate formula. Pumping a well at a constant rate and measuring the water level changes and discharge rates will provide important data which, when analyzed correctly, will describe the site-specific hydraulic characteristics of the aquifer such as hydraulic conductivity and transmissivity.

An aquifer's response to withdrawals from wells depends on its particular geologic and hydraulic characteristics. In all cases, when withdrawals start, the water levels in the well decline as water is removed from storage in the well casing. The head or water level in the well falls below the water level in the surrounding aquifer. As a result, water begins to move from the aquifer into the well. As pumping continues, the water levels in the well continue to decline while the rate of flow into the well continues to increase until the rate of inflow equals the rate of withdrawal. This movement of water toward and into the well from the surrounding aquifer results in the formation of a cone of depression.

PRELIMINARY STUDIES

Knowledge of the hydraulic characteristics of an aquifer is necessary for solving many ground-water flow problems. The problem may be local, such as determining the drawdown in a new well field, or more regional in scope, such as estimating the safe withdrawal from a ground-water basin.

Before the aquifer test can be performed, the geologic and hydrologic conditions of the aquifer must be studied. The conditions include aquifer thickness, areal extent, and location of aquifer boundaries. An aquifer boundary can be a rock-type change to one that is less permeable than the aquifer. This condition can be caused by a fault or simply result from a lateral change in the aquifer material. Areas of recharge are aquifer boundaries. A recharge boundary exists where an aquifer is freely connected hydraulically with a perennial river, a canal, a lake, or an ocean (Kruseman and De Ridder 1970).

The presence or absence of a confining layer in the vicinity of the aquifer is the major defining factor for aquifer response to a pumping test. A confining layer is a rock layer that has lower hydraulic conductivity than the aquifer. When an aquifer is bounded on the top and bottom by confining layers, the water contained within the aquifer is confined under pressure. When no confining layer is present the atmospheric pressure changes are transmitted freely through the saturated zone and the elevation of the water table fluctuates. Drilling a well into an unconfined aquifer ideally will produce radial ground-water flow toward the well and create a cone of depression. (See chapter 2 for further discussion about creation of a cone of depression.) Estimating the diameter of this cone of depression is one of the goals of an aquifer test.

In addition to locating the relative position and nature of the hydraulic boundaries, knowledge of the water table gradients and regional ground-water flow directions is necessary for an aquifer test. Some of this information can be found by researching existing well construction records and previous test borings in the general area of the new well location. If little information is available, a successful aquifer test can still be performed in these areas. This necessary information can be obtained by drilling some additional exploration wells. The basic premise behind an aquifer test is to analyze the change, over time, of the water levels in the production well and nearby observation wells that are affected by the withdrawal of water from the production well. In addition, the volume of water discharged over a given period of time is recorded and later analyzed as part of the pumping test.

KINDS OF TESTS

Particular aquifer test schemes illustrate certain hydraulic settings better than others under specific geologic conditions, so selection of the correct testing procedure will yield the most accurate results and efficient use of time. There are several varieties of aquifer tests; some involve the withdrawal of water and others analyze the changes in the ground-water table level caused by an introduction of water into the well. Each technique is testing for the same parameters, primarily hydraulic conductivity, but employs a different direction of water movement. Particular types of aquifer tests illustrate certain hydraulic characteristics depending on the type of aquifer that is tested.

A *multiple-step drawdown test* involves pumping a well at several different rates through the test. The pumping rate is increased by equal increments, each rate being held constant until a pumping level is established. At the conclusion of each step, the pump is turned off and the water levels in the well are allowed to recover for one hour. Pumping is then resumed at a higher rate. This procedure is repeated as many times as necessary until the maximum sustainable rate of pumping is obtained. The multiple-step drawdown test is used when a well penetrates an unconfined aquifer. Water in an unconfined aquifer is only under atmospheric pressure and is able to establish a new water table elevation because of the withdrawal of water.

The test is used to ascertain the performance of the well and the pump. After the results of the test are analyzed for aquifer parameters, the efficiency and operating conditions of the pump can be gleaned from the test results. In wells constructed in consolidated rock, the multiple-step drawdown test is used to determine the depth of water entry and the lowest safe pumping level.

The *constant rate test* is performed to determine the coefficients of transmissivity and storativity, the two most important aquifer parameters. This test can be run on both confined and unconfined aquifers. The effects of pumping the production well are then observed in one or more observation wells located at known distances from the test well. Since observation wells are used in a constant rate test, additional information about the aquifer can be gathered. The maximum safe capacity of a well or well field can be predicted in addition to estimating the well radius of influence. Spacing of additional wells can be determined from this test.

Much literature exists on constant rate tests. For further discussion of procedures for execution, see the article by Goodrich (1985).

The *slug test* is accomplished by injecting or withdrawing a known volume of water from a well and measuring the response of the aquifer. Depth to water measurements are performed frequently during this test (such as every minute), and the calibration of measurements is very precise because slight changes can be significant. Measurements continue until the water level has fallen or risen a significant amount, at least a few feet. This test is used primarily in situations where the hydraulic conductivity of the aquifer is low. The situation is especially likely when areas are studied for potential waste storage or disposal. The aquifer material in the area may have transmissivity values too low to conduct a pumping test.

The *recovery test*, usually performed after a pumping test has terminated, measures the recovery of the water level in the well once the pumping has stopped. Water-level recovery to the original static water level is recorded just like a pump test. The data can be used as a check on pump test data. Time and the depth of the water level are immediately recorded after pumping has stopped. The same procedure and time pattern is followed in a recovery test as in a pumping test, the only difference being the direction of change of hydraulic head. Recovery of water level usually will not return to the original static water level within a reasonable length of time, so when several measurements taken at one-hour intervals show less than 0.1 foot difference, the test is terminated.

RUNNING THE TEST

There are three basic phases involved in an aquifer test: planning, observation, and data analysis. Time and effort spent on each will produce accurate test results that can estimate the hydraulic properties of the aquifer.

Phase One: Planning

The planning phase of the aquifer test incorporates the actual organization of the mechanical aspects of the test and data collection. The planning phase is the most important part of an aquifer test and yet it is often overlooked. Extra effort and time spent on planning the test will improve the probability of accurate results and keep the costs of running the test to a minimum.

Information that is required in the planning phase includes knowledge of the geologic and hydraulic settings of the ground-water flow system, the correct type of equipment that is necessary for observation of the aquifer response, and the physical set-up of the test. Planning is important for keeping waste to a minimum. As an example, by using geologic information such as areal extent and thickness, the aquifer response can be predicted and thus the timing of measurements can be predicted. So if drawdown in a given observation well is expected to be measurable after two days of pumping, then there would be little use for water-level measurements taken at 1/2 hour intervals once the pump test begins.

Another concern that should be addressed during the planning phase is nearby production wells. These wells should be checked to determine if they will be pumping while the aquifer test is running. If nearby wells are pumping during the test, the cones of depression may intersect. If that is the case, the aquifer test results will not be a true indication of the hydraulic conditions present in the area. Other production wells should be shut down for the duration of the test or possibly pumped at a known constant rate. Correcting pumping test data for drawdown effects of other wells is difficult and should be avoided if possible.

Careful and complete data collection procedures should be addressed within the planning phase. Accuracy in every observation is vital to the success of the test. A fixed reference point for water-level measurements should be selected, marked, and made permanent for the duration of the test. The reference point can be the top of the casing, the well cap, or a place inside an access port. The reference point should be surveyed and the elevation recorded. This data point is the crux of the aquifer test. A fluctuating reference point will skew the test data and subsequently the test will not produce correct results.

A reliable backup power supply might be necessary to run the observation equipment. If a pumping test is terminated because of mechanical failure prior to the planned time, the data may not be sufficient and the whole testing effort will have to be repeated. The pump that will be used for the pump test should be of sufficient size to accomplish the task. Ideally, the pump should be run at 75 percent of capacity throughout the duration of the test. The associated equipment such as piping, wiring, and power must be reliable.

The equipment needs to be tested before the actual test date to check general operating conditions. The best rate of discharge can be approximated as well as checking the discharge measuring equipment. At the same time, changes in water levels in observation wells will be determined to assure a measurable and accurate response. All of the testing equipment should be field ready before the actual test date. Moving parts must be lubricated and spares should be available in case of breakdown during the test.

The water delivered by the well should be prevented from re-entering the tested aquifer. This can be accomplished by piping the water to an area that is not in hydraulic connection with the tested aquifer. Local recharge by infiltration of discharged water will invalidate the test. Accordingly, aquifer tests should not be run when there is likelihood of heavy rains. Proximity to high levels of infiltration or natural recharge will affect the test results and produce high values.

By using wells that are already installed, the cost of the test can be significantly reduced. Not all existing wells will be suitable for observation purposes, but close examination of each well's construction details will ultimately determine its usefulness for the test. If existing wells can be used for observation purposes, it is essential to know where the well is open within the aquifer. The drawdown data will be affected by where the existing well is open.

Phase Two: Observation

The observation phase of an aquifer test not only involves using specialized equipment, but also understanding the dynamics of the test. The fundamentals of a single well-pumping test can be applied to tests involving multiple wells. After gathering as much information as possible about the aquifer (formations encountered and when water is first observed), setting up and executing the observations necessary is the next step.

The pump selected for the job should be equipped with a control valve that will allow the pumping rate to be adjusted. This is the best type of control because pumping speed fluctuations due to line voltage changes or fuel mixture will be minimized when working against the back pressure or head developed in a partially closed valve. Changing the pumping rate by altering the piston speed is not a reliable method.

Measuring the amount of water discharged during an aquifer test is one of the primary parameters of the test. The discharged water can be measured through a variety of means, depending on the expected volume. Small pumping rates can be measured by filling a container of known volume in a measured period of time. This method is generally useful for flows of 50 gallons per minute or less. For larger flows, the common method is to use a circular orifice weir or a notched weir or to fit a water meter on the discharge pipe.

Measuring the Quantity of Water

The fundamental method of measuring the flow rate of a steady stream of water is by direct measurement of volume and weight. This is better known as the *volumetric method*. The pumped water is collected in a container of known volume in a given length of time. Variable output can be evened out by this method. The container can be almost anything, usually a 55 gallon drum or an orifice bucket. For larger quantities of water other methods are employed based on the behavior of flowing water under certain restrictive conditions. These methods involve the use of weirs, current meters, flow gauges, and similar devices.

The device most commonly used in the field is the circular orifice weir. It is compact, easily installed, and reasonably accurate. The circular orifice weir consists of a circular steel plate, about 1/16 inch thick, which is centered over the end of the discharge pipe, in which there is a perfectly circular hole machined in the middle of the plate. The diameter of the drilled hole is smaller than the diameter of the discharge pipe, and the edges of the machined hole are sharp, clean, and free of burrs. Two feet behind this plate, a small (1/8 inch diameter) hole is tapped at a right angle into the discharge pipe, and a 1/8-inch nipple is screwed into the hole. The nipple connection should be flush with the inside of the discharge pipe. Any burrs created when making the tap should be filed off.

A small, flexible, clear tubing is then connected to the nipple. This small tube is called the piezometer tube and it measures the head of water discharging through the discharge pipe. The piezometer tube consists of about five feet of transparent rubber/flexible hose attached to the 1/8-inch nipple, and the top of the tube is connected to a short section of glass tube. The water level in the tube is kept visible by raising and lowering the glass end of the flexible tube.

To measure the head in the discharge pipe, the glass tube should be held vertically next to a vertical scale that is mounted on the discharge pipe (fig. 7-1). The height of the glass tube is measured from the center of the discharge pipe to the height just above the point where the water would overflow from the tube. There are published standard tables that give the flow in gallons per minute for various combinations of discharge pipe orifice diameters (table 7-1).

There are strict limitations to this method. Certain features have to be incorporated in order for the circular orifice weir to estimate accurately the discharge rate. The discharge pipe should be at least four feet long and be supported in a perfectly level position. The discharged water has to flow freely from the pipe. The discharge valve used for controlling pump discharge should be located at least 10 pipe diameters away from the piezometer tube connection. The turbulence caused by the valve will have no effect on the orifice if it is located at this distance.

The orifice must be vertical and centered on the discharge pipe. The edges of the machined circular hole and orifice diameter must be such that the head will be at least three times the diameter of the orifice. The piezometer tube connection must be flush inside the discharge pipe. Measurement readings should be taken when the tube is free of air bubbles. This can easily be accomplished by letting water flow out of the tube between readings.

Measuring Water Levels

An accurate means to measure the water level in the well during the pumping test must be available. There are several ways to accomplish this task. Each method offers varying degrees of accuracy and dependability. The first step is to establish a fixed measuring point. To aid in quick identification it is best to actually paint the reference point so there will be no mistakes. The three most common and traditional

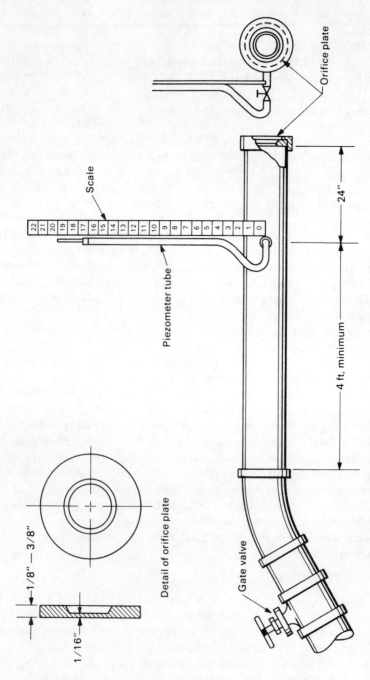

7-1. Essential details of the circular orifice weir commonly used for measuring pumping rates by means of a turbine pump. Discharge pipe must be level. (Water Well Journal Publishing Company, Columbus, Ohio)

Table 7-1. Flow Rates through Circular Orifice Weirs.

Head of Water in Tube above Center of Orifice (inches)	4-inch Pipe, 2 1/2-inch Opening (gpm)	4-inch Pipe, 3-inch Opening (gpm)	6-inch Pipe, 3-inch Opening (gpm)	6-inch Pipe, 4-inch Opening (gpm)	8-inch Pipe, 4-inch Opening (gpm)	8-inch Pipe, 5-inch Opening (gpm)	8-inch Pipe, 6-inch Opening (gpm)	10-inch Pipe, 6-inch Opening (gpm)	10-inch Pipe, 7-inch Opening (gpm)	10-inch Pipe, 8-inch Opening (gpm)
5	55	89	—	—	—	—	—	—	—	—
6	60	97	82	158	144	240	390	—	—	—
7	65	105	88	171	156	260	420	370	540	830
8	69	112	94	182	166	275	450	395	580	880
9	73	119	100	193	176	295	475	420	610	940
10	77	126	106	204	186	310	500	440	640	990
12	85	138	115	223	205	340	550	480	700	1080
14	92	149	125	241	220	365	595	520	760	1170
16	98	159	132	258	235	390	635	555	810	1250
18	104	168	140	273	250	415	675	590	860	1330
20	110	178	150	288	265	440	710	620	910	1400
22	115	186	158	302	275	460	745	650	950	1470
25	122	198	168	322	295	490	795	690	1020	1560
30	134	217	182	353	325	540	870	760	1120	1710
35	145	235	198	380	355	580	940	820	1210	1850
40	155	251	210	405	370	620	1000	880	1290	1980
45	164	267	223	430	395	660	1060	930	1370	—
50	173	280	235	455	415	690	1120	980	1440	—
60	190	310	260	500	455	760	1230	1080	1580	—

methods are wetted tape, electrical tape, and airline (fig. 7-2). Technological advances have facilitated other methods including the use of sonic sound signals and pressure transducers.

The wetted tape method is the most common of the three traditional methods. A calibrated steel tape is extended down the hole with a weight attached to the lower end to prevent any kinks or slack in the line. The lower two to three feet of the tape is coated with chalk or water-activated paste. The tape is then lowered into the well until a part of the coated section is below the water level in the well. The even foot mark is noted at the measuring point and recorded. The tape is then withdrawn and the distance between the top end and the top of the wetted end is recorded. The depth of the water level is determined by subtracting the length of wet tape from the total length of tape that was lowered into the well.

The major disadvantage of this method is that the approximate depth to water has to be known so the coated part of the line will be submerged each time. Water depths beyond eighty to ninety feet will be difficult to measure by this method because the amount of tape to withdraw each time becomes too cumbersome. The accuracy of this method is enhanced by coating the lowered portion of the tape with a water-activated paste. One such product goes on as a gold color and when in contact with water, the paste will turn red. The stark contrast of colors makes the demarcation easier to see. When there is cascading water in the well—a condition sometimes found in a pumped rock

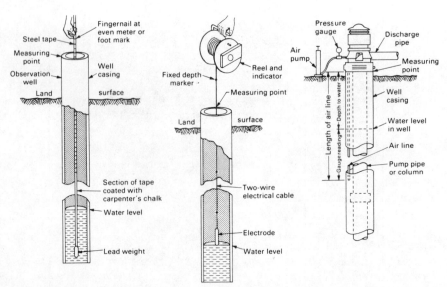

7-2. Methods for measuring the depth-to-water level in wells. (Water Well Journal Publishing Company, Columbus, Ohio)

well—the water running down the sides of the casing may mask the true mark of the water level.

The electrical tape method involves an electrode suspended by a pair of insulated wires hooked to an ammeter that indicates a closed circuit and flow of current when the electrode touches the water surface. Flashlight batteries are usually the source of power. The insulated wires are marked at one- to five-foot intervals. When the light or sounder indicates contact with water, the marked wire is held between the thumb and forefingers at the place next to the measuring point, and then more of the tape is withdrawn until the next marker appears. The distance between the held point and the next marked distance on the tape is added to the total depth. The best feature of this method is that it does not require the complete withdrawal of the tape each time for measurement; the line can remain in the well between each successive measurement. The disadvantage is the possibility of a false reading when the end is in contact with the wet well casing. If there is any oil in the well, the electric sounder may give a false reading because oil insulates the contacts. Also any nick in the wire will cause a short and give false readings.

The airline method is used for wells in which pumps will be installed, in contrast to observation wells. This method involves installation of an airline consisting of a small-diameter pipe or tube. The airline will extend from the top of the well to a point located about ten feet below the lowest anticipated water level to be reached during extended pumping conditions. The exact length of airline put into the well must be known and the submerged end of the line must be open or the method is useless.

The airline method is based on the principle that the air pressure required to push all of the water out of the submerged portion of the airline will equal the water pressure of a column of water of that same height. The pressure gauge is calibrated in feet of water to aid in calculating depth of water. When all of the water has been forced out of the line, the pressure gauge stabilizes and indicates original water column length. If the gauge reading is subtracted from the known length of airline, the remainder is the depth to water.

Other Methods of Measurement

Water-level measurements taken immediately after pumping has started or stopped are greatly affected by conditions in and adjacent to the well. Water that is stored in the casing of the well will be the first water to be pumped out during an aquifer test. If this relationship is overlooked, the test results initially will be based on the water stored in the casing.

The need to accurately record the early transition between casing storage and aquifer storage already has been addressed by the petroleum industry and subsequently modified by the water-well industry. The Johnson Research and Development Department assembled a prototype of a portable yet accurate water-level measuring device that is sensitive enough to record this transition.

The Johnson prototype weighs about 60 pounds, has a drawdown range of 99.9 feet, and is accurate to 0.1 inches. The best feature of this model is the ability of the device to measure time accurately to one second and permanently print this information on a strip of paper. This device uses a transducer that immediately transforms the water-level readings in the well into usable data. Since the diameter of the transducer is about 1 5/8 inches, water levels in small 2-inch diameter monitor wells can be measured with this device.

Another advance in measuring techniques involves sonic signals. One device in particular is made by Earth Science Instruments and is a self-contained, acoustic water-level measuring and recording system. The device can measure water levels in wells ranging from 1/2 to 2 inches in diameter. It is accurate to 0.01 feet and can be set up for unattended battery operations. The unit prints the data on paper for immediate and future use. One disadvantage of this technique is that the initial water-level reading has to be at least 500 feet for the system to work properly. The limits to the application of the system are obvious.

Test Procedures

Water-level readings in the well start immediately after pumping commences. The first reading will be after one minute of pumping time has elapsed. The depth to water is recorded with the time of measurement; the actual drawdown figure can be computed later. The most important part of this phase of data collection is to record the exact time of each measurement.

Water-level readings continue along a schedule such as ten readings per log cycle of time. Most of the readings should occur within the first few minutes of the test. If the 2-minute measurement is taken 15 seconds later, record it as 2:15. This slight deviation from the exact minute reading can be plotted on a log cycle. Late readings that are plotted as on-time readings can skew the results of the test. The duration of the test can last from several hours to several weeks depending on the type and condition of the aquifer that is being tested.

Water samples from the well are required to be taken at the end of the aquifer test. Periodic samples can be taken throughout the test and chemically analyzed to watch for any chemical or bacterial trends that might develop as pumping continues. The exact sampling procedure to be followed is outlined in the latest edition of *Manual of Methods for Chemical Analysis of Water and Wastes*. This is a U.S. Environmental Protection Agency publication that is periodically updated due to technological advances in sampling and analysis techniques.

Certain water quality parameters analyzable in the field can aid in description of the aquifer. Among these parameters are water temperature, pH, and some dissolved gasses. An aquifer test performed on a single well is a valuable tool for determining several aquifer characteristics. Tests done on a single well can determine a transmissivity value for the aquifer. Specific capacity of the well also can be derived by dividing the well yield by the maximum drawdown. This figure can be used as an index of the capacity of the well.

Aquifer Tests with Observation Wells

When an aquifer test is performed using the production well and observation wells, additional information can be gathered to further describe the aquifer conditions. A single observation well or several observation wells can be used to enhance the description of the aquifer. Each combination will yield different types of data. Some conditions warrant using observation wells whereas others may not. This concern should be addressed in the planning phase. The particulars describing one observation well can be applied to uses of several observation wells. Any observation well can be drilled specifically for that purpose or it can be a properly designed test hole or even an existing well.

A well constructed specially for an aquifer test will incorporate the ideal design features to measure the average head in the aquifer at that location. Screen placement is exactly where it should be. In confined aquifer conditions, the screen should be placed about mid-depth in the aquifer. While in unconfined aquifer conditions, the screen should be placed about one-third of the distance from the static water-level reading to the bottom of the aquifer. Construction of observation wells can be the most expensive part of any aquifer test.

A test hole can be used for an observation well when it is designed for that particular future use. Using good materials in construction of the test hole will produce a reasonably good well that has the desired yield efficiency. If the test hole is to be abandoned, inexpensive construction materials should be used. A little pre-planning can save money by constructing test holes with the capacity for aquifer response observation. For an existing well to be used as an observation well, all construction information has to be known, especially the length of the screen exposed to the aquifer.

Generally speaking, all observation wells should have a rapid response to any changes in the water table. An easy way to check response is to perform a falling head test. This test involves pouring water into the well and watching the induced head drain away. If drainage takes place in a short period of time, such as two or three hours, the well will show an adequate response. If the water does not drain within a reasonable amount of time, attempts to unclog the well should be made.

Observation well diameters need to permit rapid and accurate water-level measurements. Well diameters can range from 2 to 6 inches. Larger wells may cause a lag in response time due to water held in casing storage, so smaller wells work the best. Water-level measurements in observation wells should be done on the same schedule as the pumping well measurements. Once again, it is important to record the exact time of each measurement.

When pumping has stopped in the production well, water-level recovery measurements should be made in the observation wells. The recovery data is gathered as follows. The difference between the water level at the end of pumping and at a given time since pumping stopped is plotted as a function of time since pumping stopped. This data can be more useful than drawdown data in cases when the rate of pumping is extremely variable during the drawdown phase of the test or if the well shuts down during the test.

Placement of Observation Wells

The location of an observation well to a pumping well is a major consideration in an aquifer test. If it is too close, the water-level readings will not reflect the true hydraulic conditions. Generally speaking, observation wells in unconfined or water table aquifers need to be located closer than observation wells in confined aquifers. The distance should equal about three to five times the aquifer thickness away from the pumping well. This works out to be about 100 to 300 feet in most cases.

The number of observation wells needed for an aquifer test primarily depends on the amount of aquifer information necessary and the funds available to gather this type of information. For instance, a single observation well can yield data that when analyzed will give average transmissivity. Three or more observation wells can be analyzed by other methods and enhance the dependability of the entire analysis.

Phase Three: Data Analysis

In data analysis field observations are transformed into estimates of hydraulic properties. There are several equations that can be applied to determine the remaining unknown aquifer characteristics. Certain equations are used for specific hydraulic settings and the type of data available.

There are two ways to graphically represent the water-level changes on paper. One way is to construct a time drawdown graph. This is a plot of the drawdown against the log of time since pumping began. It is usually made on semilog paper with the drawdown plotted on the arithmetic or vertical scale and time on the log scale. A graph should be made for each observation well along with the pumping well. The other graphic representation is a distance drawdown graph. This is a plot of drawdown occurring simultaneously in each observation well against the log of the distance from each of the observation wells to the pumping well. The time selected for the graph is usually the longest available or the last reading before any interference is encountered.

The points plotted for the time drawdown graph for each well will initially fall on a curve, which in time will become a straight line within the limits of plotting. Aquifer tests performed in a confined aquifer will produce the straight line plot more rapidly than an unconfined aquifer. Extension of the straight line will approximate measured readings for continued pumping. The straight line plot will be obtained for the pumped well first. When three or more drawdown readings taken at one-hour intervals at the most distant well fall on a line, the time drawdown condition has been met and pumping can cease.

When three or more observation wells are used for an aquifer test, the distance drawdown graph is used as a check before pumping is stopped. If pumping has continued long enough, the data points will fall on a straight line within the limits of the graph. When this straight line condition is met, pumping can stop.

8

Water-Well Development

A water well is more than a hole in the ground; it is a sophisticated, engineered system for ground-water recovery. Efficient water wells provide the easiest possible path for ground water to travel from the aquifer into the well. This can be accomplished through the use of good well design and construction practices. Well-development procedures are the final step performed by the engineer and the driller.

The well-development procedure includes all steps necessary to remove fine material from the drilled borehole, correct the damage to the aquifer that occurred during drilling, and disinfect the well prior to putting it into service. The result of development is improved water passageways into the well from the aquifer, thereby maximizing the capacity and efficiency of the well.

Wells that are not developed fully will produce less water at a higher pumping cost than fully developed wells. A poorly developed well is inefficient due to the clogged pore spaces caused by the drilling process. Ground water loses energy as it flows through the damaged aquifer. The proper development procedures correct this type of damage, increasing the overall efficiency of the well system.

Engineers, geologists, and geotechnical designers of water wells must have a thorough understanding of development procedures. By understanding these procedures the designer can recommend a development method that meets the needs of the well and aquifer system. As with all design principles, no one method or procedure is applicable to all geologic and economic situations. For example, wells completed in consolidated rock aquifers require a different development procedure than wells completed in unconsolidated material. Wells that are drilled using the mud rotary method may require a different approach to development than wells drilled by the cable-tool method. A gravel-packed well requires a completely different development procedure than a naturally developed well.

IMPORTANCE OF WELL DEVELOPMENT

Every method of drilling causes some kind of damage by clogging the pores of the aquifer around the borehole. Fine material is forced back into the formation as a result of the invasion of drilling fluid during the drilling process. The wall of the borehole is sealed by the fine colloidal material that circulates through the drilling equipment. The result of this invasion of fine material into the aquifer (also known as the skin effect) reduces the porosity and permeability of the formation.

The sealing off of the borehole wall with the fine materials circulating through the drilling equipment is a necessary part of the drilling procedure. Once drilling is completed, removal of these fine materials is essential for any well to operate efficiently. The skin effect and the use of drilling fluids during the drilling process are fully discussed in chapter 3.

The objectives of development are threefold. The first objective is to clear the fine material from the face of the borehole. Second, the damaged aquifer is cleaned. Third, the fine material is removed from the well.

All development procedures have one common element. The aquifer and borehole wall are subjected to high levels of energy in order to dislodge and remove the materials that clog the pore spaces. The use of this concentrated energy to remove fine materials will rearrange the unconsolidated formation materials, forming a filtering zone in the immediate vicinity of the well screen. Creating this filter zone prevents the finer aquifer material from entering the well system and is more important than the actual grain size of the aquifer material itself. This aspect of well development is often overlooked.

Creating the filter zone not only prevents fine material from entering the well but also grades the aquifer material uniformly from coarser to finer-sized material, with the coarsest material located closest to the well. The uniformly graded material will yield more ground water than the nonuniform, heterogeneous material that existed before development. As ground water flows toward a well, its velocity increases, thus allowing ground water to flow into the well more easily.

To Correct Damage from Drilling

In all drilling operations, fine material is forced into the aquifer causing the adjacent pore spaces to be clogged. This condition is most common in mud rotary drilling, which uses drilling fluid. Other drilling methods cause similar damage to the aquifer in varying degrees. A complete discussion of mud rotary drilling and other drilling methods can be found in chapter 3. In addition to aquifer damage from the drilling fluid, the borehole wall is compressed and compacted by the drill bit. Clay and other fine-grained material encountered during drilling are smeared on the borehole wall.

Drilling of consolidated rock wells causes the plugging of fractures and crevices in hard-rock formations. Chips and particles of rock mix with water or air to form a slurry during the drilling process. The action of the drill bit forces this slurry into the openings of the rock, sealing the borehole. Proper development methods will remove these fine-grained materials from the aquifer, loosen compacted aquifer material, and clear fractures and crevices in consolidated aquifers.

To Remove Fines from the Aquifer

All aquifers contain small amounts of naturally occurring clay-sized particles. During development, it is essential to remove these particles from the area immediately around the borehole, thus preventing their entry into the well at a future date. Failure to select the appropriate development procedure results in insufficient removal of all the fine-grained material from the adjacent aquifer. This condition will result in the continual pumping of small amounts of fine-grained material along with the water (causing pump damage), migration of the particles toward the well screen (blocking and clogging the filter zone and screen slots), or both.

To Increase Local Permeability and Porosity

Proper development of a water well will create a zone of increased permeability and porosity immediately adjacent to the well. A poorly developed well will impede water movement into the well, thus creating an inefficient well system. Pumping equipment and pumping costs will increase due to the inefficient nature of a poorly developed well.

To Stabilize the Formation

The drilling process creates a borehole that is larger than the outside diameter of the screen and casing. During development, the aquifer materials are allowed to collapse around the screen, filling the annular space between the borehole wall and the screen. This movement of the once compacted aquifer helps create the filter zone around the well screen.

To Create a Filter Pack

The most important aspect of developing a water well is the creation of a filter pack around the well screen. The filter pack keeps fine-grained materials in the aquifer from entering the well. Creating this filter pack is the ultimate goal of well development. Without it, the well will produce varying amounts of fine-grained materials along with the water, causing damage to pumping equipment and damaging the well itself.

PRINCIPLES OF DEVELOPMENT

Well development can be accomplished in several different ways. Each method has its merits. In order to maximize the efficiency of development, selection of the correct method is important. A thorough understanding of the various objectives behind development is necessary before an educated choice can be made.

At first glance the objectives of development may appear simple. Field experience has shown that this essential phase of well construction is often delegated to the contractor instead of the well designer. Although the contractor may be well versed in one or more development techniques, the well designer must be consulted in this and other decisions during the actual development procedure. Ideally, the contractor and well designer should work together, combining their knowledge and expertise. Developing a water well is a complex series of tasks, beginning with the selection process. Aquifer characteristics such as grain-size distribution and well construction details are the primary criteria used for selecting the well development method.

The first objective of development is to disturb the aquifer material with some form of energy sufficient to cause the material to collapse around the well screen, essentially filling the annular space between the borehole wall and the well screen and casing. Disturbing the aquifer materials is basically done to correct damage to the aquifer due to the drilling process. During the drilling process the aquifer is compacted and sealed off with clays and other fine-grained materials used in drilling fluids. If the borehole wall is not disturbed, these fine-grained materials may be forced further back into the aquifer, thereby making development of the well even more difficult.

Selecting a method that will attack the borehole wall is absolutely essential when dealing with gravel-packed wells. In this case, an envelope of gravel or sand has been placed between the outside of the screen and the face of the borehole wall. If the energy used during development is not sufficient, only the gravel pack will be developed and not the aquifer.

During the initial phase of development, the energy intensity of the selected procedure should be slow and gentle. If the well is being naturally developed, the aquifer material will collapse around the screen and casing. Applying too much intensity at this point could result in the entire aquifer collapsing all at once, placing a tremendous amount of stress on the screen and casing. Using a method that initially requires too much energy may result in collapse of the casing and/or the screen, rendering the well useless.

Once the aquifer has collapsed around the screen and casing in a controlled manner, the intensity of the development procedure can be increased. In gravel-packed wells the same philosophy regarding the intensity of the development energy holds true. Always start development using low-energy procedures, followed by increasingly intensive procedures.

The development process also must disperse the fine colloidal particles that were used to seal the borehole wall during drilling. Removal of these dispersed particles in a timely fashion is just as important as selecting the correct development procedure. Removing these fine particles from the well system during development eliminates the possibility of movement back into the aquifer.

By disturbing the aquifer materials, the fine particles are liberated more easily. These particles may be dispersed automatically when the development procedure uses water as the source of energy. Chemical additives such as phosphates and surfactants enhance this dispersing action by reducing the electric charges that hold clay molecules together. When these charges are reduced, the clay particles are free to move, allowing them to be easily removed by discharging water from the well.

Automatic dispersion of clay particles is not common when air is the energy source used to develop a well. Because air is a convenient energy source common to many of today's drilling rigs, many contractors prefer to use it. Steps must be taken to ensure that the initial phase of development (disturbing the aquifer materials) includes proper dispersion techniques.

Once these fines are dispersed, they must be removed from the well structure. If the fine materials are allowed to stay in the well, they will move back and forth from the aquifer and into the well during development. Once development stops, a portion of these particles will remain in the aquifer, blocking and clogging the aquifer pore spaces. Removal of these fine particles may be somewhat difficult in small-diameter wells, but is easily accomplished in larger diameter wells by using an air compressor to displace standing water from the well. Water from the aquifer is encouraged to enter the well system, creating water passages from the aquifer to the well. When the well diameter is large enough, the well can be pumped using compressed air while simultaneously developing the well.

Well development must direct energy into the aquifer or the gravel pack. Sufficient energy is needed so that the adjacent materials will grade according to size, creating a filter zone around the well screen. The development procedure that is finally chosen must have enough available energy to accomplish all of the aforementioned objectives. Necessary energy levels vary between development methods. Consequently, under the correct conditions all development methods can be successful. In all cases, the development procedure must be to create the filter pack around the well screen either from the aquifer materials or from the installed gravel pack.

Finally, the energy source used for the development procedure must be depth-specific within the well, concentrating development in the zones that show poor performance. The ability to selectively develop one section or zone of an aquifer is a desirable characteristic for selection of a development method. Wells with long sections of screen dissipate development energy wherever the easiest path into and out of the aquifer exists. The ability to intensify the development procedure in selected areas will allow for the entire screened interval to be developed, thereby increasing the overall efficiency of the well.

WELL-DEVELOPMENT PROCEDURES

There is no single, effective well-development method; the effectiveness of each method employed depends on several factors including:
- local geologic conditions
- the drilling method used to construct the well
- design of the well (including such factors as the screen length and type, the diameter of the well, and whether the well was gravel-packed or naturally developed)
- availability of equipment

There are numerous development methods available. Some of these methods are more effective than others, while some are used for specific purposes and results. In order from least to most effective, the most common development methods include:
- overpumping
- rawhiding
- surging with a surge block
- air lifting/pumping
- air surging
- high-velocity jetting

Selecting a development method for a particular well depends on many factors. Most of the development methods listed here are used in combination with each other. As stated in the previous section, the development procedure must begin slowly, so as not to collapse the aquifer materials around the screen intake area all at once. Therefore, even the least effective methods have their place.

Developing a water well is more an art than an exact science. Experienced contractors and designers have developed an intuitive sense of what is needed to get a well into service. Some aquifers respond better to one particular development method, and other aquifers require a longer development time. One of the best sources of predesign information about development methods in a particular area is the local contractor. A predesign meeting with one or more local contractors may eliminate cost overruns during the development phase of the contract.

Overpumping

One of the simplest methods of developing a well is to overpump the well. This can be accomplished using a temporary pump that is installed in the well, or with compressed air, which is generally available on the rig. Overpumping the well is not an effective development method, but it can be used to initiate the development procedure by removing some of the drilling fluids and other fine material from the well.

Some contractors have pumps that are used solely for overpumping, as this method erodes the impellers of the pump rapidly. Other contractors prefer to use compressed air to pump the well. However the well is pumped, it is essential to start the procedure slowly, increasing the volume of pumping with time. As the procedure continues, the pumping rate of the well is increased until the rate of pumping exceeds the expected production rate of the well.

Overpumping of a well does not accomplish any of the development principles that were discussed in the previous section. Overpumping the well does not agitate the aquifer, nor does it provide for the necessary motion of water into and out of the aquifer. It only supplies a one-way action of water movement that pulls fine material toward the well and allows the aquifer to develop the least resistant water-flow pathways.

Sand bridging in the aquifer commonly occurs when overpumping. Sand grains pack around themselves in response to the one-way direction of water movement. Water that is being pumped from the well appears to be clear and free of fine materials. Once pumping has ceased, the sand bridges break down, becoming loose particles of sand again. When pumping resumes, the initial volume of water that is pumped from the well is laden with the fine sand grains that are closest to the screened area. The unidirectional motion of water will again create sand bridges, starting the process again.

Overpumping, despite its poor results, can be used as part of the overall development procedure. During the initial phase of development, overpumping of a well will help clean out the fine materials and begin to collapse the aquifer material around the well. It is important not to pump the well excessively at this stage. Overpumping also is used during the final step in the development procedure. Pumping the well excessively and monitoring the well's yield will establish a base well yield that the contractor and designer can work from when the well is pump tested.

Overpumping is used in wells that will be shut off rarely. If the desired yield from the well is obtained during the initial overpumping and the well is not going to be turned on and off repeatedly, overpumping may be the only development method that is needed. Sand bridges will be created in the aquifer, eliminating or at least reducing the amount of sand grains that enter the well. Wells, such as dewatering wells at construction sites, are rarely shut down until construction is completed. In this case, overpumping of a well that produces the amount of water needed to make the dewatering effort successful might be the only necessary development method.

If too much sand enters the system, further development may be needed. Small amounts of sand and fine particles will eventually erode the pump impellers or bowls to the point that the well system will fail. This could become a crucial situation if the construction is at a critical point. Monitoring of the dewatering system is important.

Rawhiding

Rawhiding a well is basically a variation of the overpumping method. As with overpumping, there are very few rules to follow. The success of the procedure is somewhat dependent on the experience of the contractor. Inexperienced contractors do not completely understand the processes at work and are not as willing to turn the pump on and off as required, but opt to let the pumping system operate for longer periods of time. This tends to defeat the purpose of rawhiding.

Rawhiding a well is a fairly straightforward procedure. All check valves, foot valves, and other backflow prevention devices are removed from the pump. The pump or the tail pipe of the pump is placed in or near the screen area. The pump is turned on, and the well is pumped until the water from the well appears clear and free of fine sand. The pump is turned off, and the water that is left standing in the discharge pipe begins to flow downward. The water passes through the pump and flows out into the well. Reversing the flow of water tends to disturb the aquifer material that is adjacent to the well screen. Sand bridges that were created during the pumping phase are broken down and the sand grains become mobile again.

The pump is started again, initially producing dirty, sand-laden water from the well. Once the water clears, the contractor turns the pump off, creating a backwash effect. To be effective, the pump must be lowered and raised in the well, backwashing the entire screened area. This can be very time consuming and costly for wells that have long sections of screen, but may be very effective for wells with short sections of screen that are constructed in clean, gravelly formations.

Although rawhiding is not the best method available, it can be used on wells that meet certain criteria. The well must have a high specific capacity. Wells with low specific capacities will not benefit because they require more drawdown than wells with higher specific capacities.

When the pumping stops and the water that is standing in the drop pipe or column pipe begins to flow back into the well, the water will simply flow into the well and help establish the static water level. It is only when the water in the well exceeds the static water level that the backwashing action occurs. Wells with high specific capacities have less drawdown, and thus less water is needed to refill the well back to the static water level. The remaining water is forced into the well, creating an artificial head and allowing water to flow back into the formation.

Wells that have a high static water level will not benefit from rawhiding. Again, the water in the drop pipe or column pipe must be of sufficient volume to create an artificial head within the well. If the static water level is high (near the surface of the ground) the drop pipe may not hold enough fluid to fill the well past the static water level. Wells that have a deep static water level would benefit due to the larger volume of water stored in the drop pipe.

Due to the expense of installing a pump and moving it up and down long sections of screen, rawhiding may prove too costly in deep wells. The wear and tear on the pump itself must be considered. Generally, pumps are made to turn in one direction; that is, pump bushings and bearings are designed to allow the pump to operate in one direction. As water flows back into the pump from the drop pipe or column pipe, the impellers in the pump are spun in reverse. The weight of the water and the reverse spinning action will cause excessive wear to the pump and motor. Additionally, if the pump is started while it is still spinning in the reverse direction, permanent damage to the pump and motor may occur.

Rawhiding a well as a method of development is of limited use. Rawhiding will not allow the well to be fully developed because the water tends to flow back into the aquifer only through the paths of least resistance. Only those zones of good production will be developed, leaving the remainder of the aquifer virtually unaffected by the development effort. The method does not penetrate the aquifer very deeply, thus making the development of a gravel-packed well virtually impossible.

Rawhiding is an improvement over the overpumping method, but can be used only in a narrow spectrum of wells. Wells that are constructed in a clean formation and that have high specific capacities and deep static water levels will benefit from the use of the rawhide method of development, but other methods may be better and more cost effective.

Surging with a Surge Block

Surging, one of the most common methods of developing a well, consists of moving a plunger-like tool up and down inside of the well casing and screen. A tight-fitting tool called a surge block is inserted into the casing of the well and is forced up and down, creating a suction action on the upstroke and a pressure action on the downstroke. This suction and pressure action develops the well screen by forcing water in and out of the intake area.

Surging is a simple method and requires little experience. The surge tool can be a homemade device consisting of a block of wood about 1 to 2 inches smaller than the inside diameter of the casing into which the surge block will be inserted. The block of wood is cut into several horizontal slices and fitted with circular pieces of rubber strapping that are about 1/2 to 1 inch larger than the diameter of the casing. On the downstroke, the rubber strapping is forced upward, sealing the gap between the block of wood and the well casing. On the upstroke, the rubber strapping is forced downward, again filling the gap between the block of wood and the inside of the casing. This action produces the suction/pressure action. Surging is similar in action to clearing a clogged drain pipe with a plunger.

There are numerous variations to the basic surge tool design. Many of these devices were invented to develop wells of a particular type and may not work in all situations. One of the most common variations is to add a flapper type of valve on the surge block. The valve is opened in one direction and is forced shut in the other. By removing the tool and turning the surge block upside down, the opening and closing action of the valve is reversed.

The valve type of surge tool is popular because it allows the water to move only in one direction, until the surge block is turned upside down, thereby reversing the direction of water flow during development. Contractors often place the tool in the well so that the valve creates a suction action on the well screen. The contractor then pushes the tool down the well, allowing the valve to open, and the surge block moves easily downward, exerting little or no pressure action. The contractor pulls up on the tool with the rig and the valve closes, creating a suction action in the well.

As the tool is pulled upward, water is produced from the aquifer, filling the void space that is left by the rising surge tool. As the tool is forced downward, water passes through the valve, and is captured above the top of the tool. When the tool is pulled upward, the valve closes and traps the water that is above the tool, creating a small pumping action within the well. This action continues for a short while, until the tool is removed and the fine material that accumulated in the well during the development process is removed. The contractor then reinserts the tool, this time in the reverse position, and forces water into the aquifer, breaking down the sand bridges that have formed in the aquifer. The tool is then removed from the well and the process begins again.

Other variations in the surge tool design include a pneumatically operated device that moves a plate up and down in the well. A stationary plate is positioned above the moving plate to help concentrate the action of the moving plate. This type of surge tool was developed for use primarily in limestone and sandstone aquifers. Another variation in surge tool design is a tool that consists of two individual surge blocks that are connected to each other by a pipe. The pipe is then connected to the surface, allowing the rig operator to force the tool up and down in the well. The section of pipe that connects the two blocks together is slotted, allowing chemicals to be introduced from the surface. This tool was originally designed for work in rehabilitating wells that have a scaling problem. Acids and other cleaning agents were introduced into the screen area, concentrating their effects to those areas that showed the most scale build up. The tool was slightly modified later and used in developing wells; dispersion agents were introduced into the well screen area to break up clay and other fine colloidal particles.

Air Lifting/Pumping

If compressed air is available on the rig, most contractors will use it as a means of developing a water well. The first tendency of contractors and designers is to use the tremendous power of compressed air to create a violent and turbulent environment in the well. This technique appears to produce beneficial results, but a closer examination will reveal that compressed air can, at times, be a poor method of development. In pumping the well, compressed air is used by placing the drilling tools — without a bit — down the well and forcing air down the drill stem and out into the well until a slug of water appears at the surface in a violent manner.

A drawback of the compressed air method is that the water in the well and aquifer is pulled into the well and pumped out at such a rapid rate that the net action or movement of the water in the aquifer is in one direction: into the well. This one-way action of water is not the desired action necessary to properly develop the well. The well is developed too quickly. Our previous discussion on the principles of development mentioned that the chosen development techniques must be based on an initial low, then progressively higher, energy level. Failure to do this may result in collapsing the well, reducing the formation permeability around the well, and perhaps permanently reducing the overall yield of the well.

Using compressed air can ruin a well by filling the aquifer up with tiny air bubbles, reducing the permeability of the aquifer and thereby reducing the yield of the well. This can occur when the contractor uses too much air, too fast, causing the air to move back into the aquifer. Once in the aquifer, the air becomes trapped in between the grains of sand and is held there due to the surface tension of the water. These air bubbles will dissipate from the aquifer in time, being dissolved in the water that passes by them.

Pumping with air should be used only in a high-production well after the development phase is nearly completed. If it must be used during the initial development phases, the amount of air used in development should be limited.

Air Surging

Air surging is a variation of both air pumping and surging a well. This method uses compressed air to force water back into the formation, and to pump the well to remove the fines and to set up water paths. Two pipes are dropped into the well, each capable of being raised or lowered into the well independently of each other.

The larger pipe, called the pumping or eductor pipe, has a diameter about two inches less than the inside of the casing. This eductor pipe will channel or carry water out of the well. The eductor pipe is sealed at the surface and fitted with a valve so that the flow of water and air can be shut off or released quickly. The contractor will begin the development procedure with the eductor pipe near the bottom of the intake of the well, with the bottom of the airline about four to ten feet inside of the eductor pipe. Nested inside of the larger eductor pipe is a smaller pipe (usually about one-half to one-fourth the diameter of the eductor pipe), which will carry the compressed air into the well.

The technique is fairly simple. The contractor begins pumping air down the smaller air pipe and allows water and air to be pumped out of the well at a slow but steady rate. Once this system has stabilized, the contractor closes the valve for a very short amount of time and lets the air and water pressure build up in the eductor pipe. The contractor quickly opens the valve, releasing the built-up air pressure and water in the eductor pipe and allowing the well to stabilize to its original pumping rate. This procedure is repeated over and over again until the well clears.

Once the well is cleared, the contractor moves the eductor pipe a few feet further up into the intake and repeats the process. In areas of high sand production, the contractor may wish to push the airline out of the bottom of the eductor and into the intake; this allows the compressed air to create more turbulence in that particular area of the intake, thereby speeding up the development procedure.

High-Velocity Jetting

Generally considered the most effective method of well development, jetting with water at high velocities has the following advantages:

1. The jetting force is concentrated over a relatively small area for optimum effectiveness.

2. The jetting tool can be directed selectively to every part of the well screen. If screen openings are designed to give maximum exposure to the surrounding formation, complete development can be achieved.
3. The jetting action is easily controlled and applied, and problems resulting from overapplication are not likely to occur.

Basically, the equipment consists of a simple jetting tool with two or more nozzles, hose, and piping that are connected to a high-pressure pump, and a water tank or other water supply. The jetting action is directed through the screen openings to effectively agitate the sand and gravel particles of the surrounding formation. The turbulence created by the water jets causes the sand and silts to re-enter the well through the screen openings above and below the tool's point of operation. When conditions permit, the well should be lightly pumped to help remove the particles and to facilitate drawing the particles into the well.

The horizontal jetting tool should be raised and lowered as well as rotated to effectively reach each portion of the well screen and thus each area of the surrounding formation. For wells drilled with the conventional rotary method, this jetting procedure effectively breaks up and disperses the filter cake remaining on the borehole wall and allows the mud mixture to be pumped out of the well.

Depending on the jetting depth and the required gallons per minute, pipe sizes vary from 1 1/2 to 2 to 3 inches in diameter. The proper size will hold any friction losses to within acceptable limits. The lowest effective velocity is about 100 feet per second, with much better results occurring at velocities of 150 to 300 feet per second. Jetting tools with either two, three, or four nozzles are evenly spaced to provide hydraulic balancing of the tool during operation.

9

Water-Well Maintenance

The deterioration of a well begins as soon as the well is put into service. The process is usually gradual and to some extent predictable. A sound preventive maintenance program for a water well has always been important, but it is especially so today with rising energy costs and higher capital costs for installing replacement wells. A maintenance program that prevents a loss of efficiency can amount to substantial money savings over the life of the well.

Deterioration of water wells with time is a function of the interaction between the well structure and the ground water. The processes of deterioration are controlled by several factors. These factors can work separately or together to accelerate the deterioration of water wells. Proper well-design practices will help alleviate or stall the severity of some of these processes.

CAUSES OF WELL FAILURE

A well may fail through incrustation, chemical corrosion, or mechanical erosion. Field experience has shown that these types of failures are usually interrelated. For example, screen slot openings may become enlarged due to the corrosive nature of the ground water. This enlargement will permit sand to pass through the screen, which will eventually erode the pumping equipment.

The most common type of well failure is caused by the deposition of encrusting materials that effectively block the intake section of the well. This process is sometimes called scaling and can be broken down further into three kinds: chemical, physical, and biological. About 90 percent of incrustation problems are caused by soluble chemicals in the ground water that precipitate out of solution and are deposited in an insoluble form in the vicinity of the well screen.

Chemical corrosion can cause a loss of strength in the casing, screen, or pump, resulting in structural failure through collapse of the well structure. Corrosion occurs as a result of two general processes: chemical action and electrochemical action.

Breakdowns in pumping equipment can occur as a result of mechanical erosion. Combinations of any of the previous types of well failure can result in pumping sand and subsequent erosion of the entire well structure and pumping system. These types of breakdowns can be avoided through properly monitoring and maintaining the well and pump. Consult the pump manufacturer for maintenance details.

TYPES OF WELL FAILURES

Incrustation

Incrustation in a water well is defined as the deposition of organic or inorganic material within the aquifer near the borehole; in the gravel pack; around the screen; upon surfaces of the screen, casing, or pump; or any combination of these. The resulting deposits restrict the water passages from the aquifer into the well and increase surface roughness. This in turn increases local ground-water velocities, causing turbulent flow and subsequently reducing the specific capacity of the well.

The process of incrustation can be divided into three main types: chemical, physical, and biological. Chemical incrustation is caused by the amount and type of minerals dissolved in the ground water. Physical incrustation is the result of fine materials clogging and filling in the well-screen openings and plugging the adjacent aquifer. Biological incrustation is the result of the well being an ideal subsurface environment in which certain strains of nonpathogenic bacteria thrive. If this condition is left unchecked, the proliferation of these types of bacteria can clog not only the well screen but the entire well structure, pump, and distribution system.

Chemical Incrustation

The type and amount of chemical incrustation is dependent on the ground-water quality coupled with the design of the well. The resulting scale is primarily composed of carbonate minerals precipitating out of solution. Iron and manganese hydroxides and hydrated oxides can contribute to the incrustation problem. Regardless of the well-screen materials, soluble minerals found in ground water will precipitate out of solution, causing chemical incrustation.

Carbonate deposition. Calcium and iron bicarbonate are carried in solution in proportion to the amount of dissolved carbon dioxide in the ground water. The ability of ground water to carry carbon dioxide varies directly with the amount of pressure in the subsurface. An increase of pressure will permit more dissolved carbon dioxide to be present; thus the calcium content of the ground water will increase.

When water is pumped from the aquifer, the water table or piezometric head declines, lowering the hydrostatic pressure exerted on the ground water; this upsets the chemical equilibrium of the ground water in that area. The most dramatic change in the ground-water equilibrium occurs at the well-screen/aquifer interface.

Because of the hydrostatic pressure drop, carbon dioxide will be released or degassed, and the soluble bicarbonate carried in the ground water will convert to the insoluble carbonate form. The insoluble carbonate is then deposited on the face of the screen, plugging the slots and reducing the yield of the well. This chemical reaction and loss of carbon dioxide is as follows:

$$Ca(HCO_3)_2 \rightarrow CaCO_3 \downarrow + CO_2 \uparrow + H_2O$$

$$Fe(HCO_3)_2 \rightarrow FeCO_3 \downarrow + CO_2 \uparrow + H_2O$$

Total energy in any water mass consists of three components: pressure, velocity, and head. The sum of the energy potentials (H) is expressed in Bernoulli's equation:

$$H = \frac{p}{\gamma} + \frac{V^2}{2g} + z,$$

where p is pressure, γ is the specific weight of water, V is the velocity of flow, g is the acceleration of gravity, and z is the elevation above a certain datum. According to this equation, pressure decreases proportionately to the square of the velocity. In a ground-water system, hydrostatic pressure decreases due to the changes in ground-water velocity. For instance, the pressure drop at 2 feet per second is four times greater than the pressure drop at 1 foot per second.

As water passes through the slots of a screen, it accelerates; thus, the pressure decreases and carbon dioxide degasses. Chemical precipitation will then occur in and around the well screen. Once the chemical precipitation begins, the screen slots begin to close, increasing the ground-water velocity through the screen even further. This cycle will continue to accelerate with time.

Iron oxides. Changes in velocity of the ground water also cause precipitation of iron oxides. In a pumped water well, the cone of depression provides a mechanism for air to enter the pore spaces of the aquifer, oxidizing and precipitating the iron dissolved in the ground water. As pumping starts and stops, a coating of iron oxide gradually develops on the mineral grains, and the volume of the pore spaces is reduced.

An additional factor that can lead to iron precipitation problems is air in the casing, which will provide a constant supply of oxygen. The precipitated iron deposits are either in the form of hydrated ferrous oxide, which is a black sludge, or ferris oxide, which is a rust-colored scale. Other forms of iron precipitates can exist.

The chemical quality of ground water is an indicator of incrustation potential. The correlation between quality and incrustation is complex; however, the principal indicators that encourage chemical incrustation are listed as follows:

1. pH greater than 7.5
2. Carbonate hardness greater than 300 mg/l
3. Manganese greater than 1 mg/l plus high pH and dissolved oxygen
4. Iron greater than 2 mg/l

Unfortunately, there is no way to prevent totally or to accurately predict the magnitude of chemical incrustation problems. Wells can be designed to incorporate features that work better with certain ground-water chemistries, extending the maximum efficiency period for the well. Timely maintenance can keep the incrustation problem in check, thus extending the overall usefulness of the well.

Physical Incrustation

The clogging of screen openings with solid particles carried from the aquifer is a common form of incrustation. This condition results primarily from the failure to effectively remove fine particles from the gravel pack and surrounding aquifer during development. Improper design of the gravel pack also can cause this form of incrustation.

Physical clogging is accelerated by improper well design. For example, slot geometry of the screen may encourage sand grains to be trapped in the slot opening, thereby decreasing the effective open area. A screen that is slightly smaller on the outside than on the inside helps to eliminate this situation. Selecting a gravel-pack size that is too large for the aquifer defeats the purpose of the gravel pack. Fine aquifer material will migrate toward the well, filling up the pore spaces of the gravel pack. This reduction in porosity increases the ground-water velocity through the remaining open passages.

Biological Incrustation

A variety of organisms are capable of deriving their nutrition and energy from ground water containing dissolved iron. Some organisms can convert iron bicarbonate to ferric ions (insoluble precipitate) and consume the carbon dioxide that is released. Others use organic compounds found in ground water as food and release ferric ions as a byproduct. Other biological organisms appear to convert ferrous ions to the ferric ion form when either organic or inorganic compounds of iron are present in ground water.

In most cases the result is the development of a complex deposit consisting of a reddish-brown hydrated iron oxide, a black iron oxide film, and a gelatinous mass of living organic matter. Both the precipitation of iron and the rapid growth of bacteria create a voluminous mass of material that quickly plugs the water passages of the adjacent aquifer, gravel pack, and well screen. The predominant strains of iron bacteria found in water wells are Siderocapsas (encapsulated rodlike cells), Gallionella (twisted bands of cellular filament), and Clonothrix and Leptothrix of the *Crenothrix* genus (multibranched sheaths of rodlike cells).

Iron bacteria favor environments with ferrous ion concentrations as low as 0.02 mg/l. However, 1 mg/l is probably a more realistic threshold. Shallow water with temperatures below 24°C, pH between 5.4 and 7.2, and dissolved solid contents below 1000 mg/l are other parameters that appear favorable for iron bacteria development. Some of the organisms, like Leptothrix, flourish only at high dissolved oxygen concentrations, whereas others are well adapted to oxygen concentrations below 5 mg/l.

Iron bacteria favor areas within the well system that create a high-flow velocity environment. Areas such as the pump intake, column pipe, and well screen are prone to development of iron bacteria colonies. Unfortunately, all of these areas seriously impact well operations.

Several theories pertain to the origin of iron bacteria. One is that the bacteria is already present in the aquifer, but it is the change of environment due to a pumping water well that causes them to proliferate. This theory has won support through projects that use naturally occurring bacteria for aquifer restoration projects. In this application, favorable environments for bacteria that are naturally occurring in the soil and aquifer are created. These bacteria are encouraged to grow and consume the contaminant as part of their life cycle. Another theory for the origin of iron bacteria asserts that iron bacteria spores are transported from well to well through drilling and pump-handling equipment. Once a colony is established, bacteria spores can be transferred between wells. When iron bacteria problems arise, they cannot be solved permanently, but they can be arrested and somewhat controlled by the use of chemicals (chiefly chlorine) and other techniques.

Sulfate-reducing bacteria is another bacterial problem for water wells. It is not as common as iron bacteria due to the type of environment this strain of bacteria favors. Sulfate-reducing bacteria is an anaerobic organism that reduces sulfate to sulfide, which in turn combines with hydrogen in water to form hydrogen sulfide. High concentrations of hydrogen sulfide in groundwater can corrode steel well parts.

Prevention of Incrustation

The following modifications to the water well system may reduce the potential for mineral and biological incrustation. These modifications can be implemented singularly or together, but may not be totally successful.

Increase the open area of the well screen. By increasing the open area of the screen, ground water enters the well more freely. Friction, head loss, and velocity of the ground water are all reduced, thereby reducing the tendency for chemical precipitation and subsequent scaling to occur. This modification can be incorporated into the design phase of the water well or be used as a rehabilitative effort. The effects of the open area of the screen on the operation of a water well can be found in chapter 5.

Improve the efficiency of the well. By improving the efficiency of the well, ground-water head loss through the well screen is reduced, which in turn reduces the tendency for chemical precipitation to occur. The efficiency of the well is improved through further development of the well. Maximizing the efficiency of a well can be incorporated in the design phase. Selecting the proper screen design and development method prior to installation will minimize incrustation tendencies over the life of the well.

Reduce the pumping rate of the well system. By spreading the pumping load over several wells, the pumping rate from an individual well is reduced, so drawdown will decrease. A decrease in the drawdown of a well reduces the head loss, friction loss, and velocity of the ground water as it enters the well. If the pumping rate of the well is not crucial to the water system, but the total yield from the well is of concern, the well can be pumped at a reduced rate for longer periods of time.

Follow a proper preventive maintenance program. By monitoring the physical and chemical changes of a well, trends in the well's performance can be obtained. Using this data, the operator of the well system can predict an incrustation problem before it occurs. A proper maintenance program can prevent the incrustation problem from becoming critical.

Reduce the oxygen entering the well screen. Bacterial growth and chemical precipitation may be dependent on the presence of oxygen. By reducing the amount of available oxygen in the ground water, bacterial growth and precipitation of metallic oxides are reduced significantly. This can be accomplished by reducing the drawdown in the well, which increases the pumping water level, preventing air from being dissolved in the ground water adjacent to the screen. Reducing the drawdown will decrease the cone of depression, which reduces the amount of oxygen entrained into the dewatered portion of the aquifer.

A drawdown seal can be installed in the well, creating a physical barrier between the atmosphere and the ground water in the well. A drawdown seal is a plate that seals itself against the casing. It is installed just above the pump and is designed to allow water from the pump to pass through it without allowing air to enter the well.

Corrosion

Corrosion is the chemical or physical decomposition of materials. There are several different types of corrosive forces that affect a water well. Chemical and galvanic corrosion and physical erosion are the most common forces that are at work in a well system.

Chemical corrosion is the process that removes metal ions from the surface of the corroding metal. This type of corrosion is thought to occur in ground waters that would not function as an electrolyte (a carrier of electric current) due to low amounts of dissolved solids. Since there are no naturally occurring ground waters with such low dissolved solids, corrosion by simple chemical action is hard to distinguish from electrochemical corrosion.

Electrochemical corrosion means that a chemical change is accompanied by the flow of electrical current. This type of corrosion process usually requires considerable salts in the ground water to make the metal an excellent electrolyte. Electrochemical corrosion has to have differences in electrical potential in order to get started. This happens when two different kinds of metals are present whereby material is removed from the metal that is lowest in the electromotive series. The electromotive series is a relative ranking of metals' potentials for electrochemical corrosion. The farther apart two metals are on the electromotive series, the larger the differences in electrical potential, thus increasing the rate of corrosion proportionately.

In electrochemical corrosion, both cathode and anode are created and metal is removed from the anode and deposited at the cathode. The following list is a modification of the electromotive series.

Corroded End
magnesium
magnesium alloys
zinc
aluminum 25
cadmium
aluminum 17 ST
steel, iron, cast iron
chromium-iron (active)
Ni-resist
18-8 stainless steel (active)
lead, tin, lead-tin solders
nickel, inconel (active)
brass, copper
bronze, Monel
silver solder
nickel, inconel (passive)
Continued

18-8 stainless steel (passive)
silver
gold, platinum
chromium-iron (passive)
Protected End

There are several water-quality parameters that can indicate possible corrosive conditions to a water well. To anticipate corrosion problems, chemical analysis of water samples should be performed at regular intervals. Listed below are several of the classic conditions that can lead to corrosion problems.

- acidic water (pH < 7.0)
- dissolved oxygen (DO) > 2 mg/l
- hydrogen sulfide (H_2S) > 1 mg/l
- total dissolved solids (TDS) $\geq$ 1000 mg/l
- carbon dioxide (CO_2) $\geq$ 50 mg/l
- chlorine (Cl) > 300 mg/l
- higher temperature increases corrosion by accelerating hydrogen evolution

By closely matching the material used in the well screen to each ground-water environment, the design life of a well screen is extended.

TROUBLESHOOTING

The performance history of the well is the starting point for determining what the problem is or where it originated. Knowledge of the current operating conditions is necessary for determination of the apparent problem. It should become obvious that keeping accurate performance records is the key to understanding and correcting well-performance problems. There are three tools that can be used to troubleshoot well problems. The first and second are historical data and actually seeing the problem. The third is a water-quality analysis of a water sample that can be used for comparison purposes.

Problem Solving by Interpreting Field Observations

Observed: Heavy reddish-brown iron oxide, stains in discharge, or red water.

Possible conditions:
Indicates anaerobic conditions (a reducing environment)
- Favorable environment for either inert or biological iron deposits

- Corrosive waters attacking metal parts in the well
- Possible iron bacteria

Indicates aerobic conditions (an oxygen-enriched environment)
- Aeration of water in well (cascading): yield/specific capacity of well has dropped; possible hole in casing; or iron oxide scaling

Course of action:
Test chemical and bacterial content of water
Perform videocamera survey of well (to see bacterial growth, scale, and/or cascading water)

Observed: Bubbles in the discharge water.

Possible conditions:
An environment with free carbon dioxide
Cascading water
Naturally dissolved gases in the water
Overpumping of the aquifer

Course of action:
Test chemical and bacterial content of water
Perform videocamera survey of well (to see bacterial growth, scale, and/or cascading water)

Observed: Rotten egg smell (hydrogen sulfide odor).

Possible conditions:
A reducing environment

Course of action:
Test chemical and bacterial content of water for hydrogen sulfide concentrations; if levels are 0.5 mg/l or greater, damage may result to copper alloy parts of the well system
Perform videocamera survey of well (to see bacterial growth and effects of corrosive environment)

Observed: Well efficiency decreased.

Possible conditions:
Chemical and/or mechanical incrustation
Biological fouling
Decrease in regional water table
Structural collapse caused by corrosion or other factors
Change in water quality
Improper well design and construction
Pumping in excess of design

Course of action:
  Test chemical and bacterial content of water
  Perform videocamera survey of well (to see bacterial growth, scale, and/or cascading water)
  Test pumping system and well
  Measure water levels after well recovers and during pumping

Observed: Reduced production; loss of pressure.

Possible conditions:
  Pumping system deteriorated
    • Consult manufacturer
  Change in pumping water level
  Reduced static level
  Decrease in well efficiency

Course of action:
  Test pumping system and well
  Measure water levels after well recovers and during pumping
  Measure pump shut-off head
  Measure wire-to-water efficiency of the pumping system
  Inspect pumping system

Observed: Sand in discharge; loss of pressure and/or land subsidence around well.

Possible conditions:
  Inadequate or improper well design
  Inadequate or improper well construction
  Well screen and casing deterioration caused by corrosion
  Well being pumped in excess of design intent

Course of action:
  Review well design
  Review well construction for conformity to design
  Measure water levels after well recovers and during pumping

Observed: Pump is cavitating, providing variable discharge, and/or breaking suction.

Possible conditions:
  Well being pumped in excess of design intent
  Drawdown level excessive in well
    • Regional water level has declined
    • Well efficiency decreased
  Well screen and casing deterioration caused by corrosion
  Well screen encrusted

Course of action:
   Test pumping system and well
   Measure water levels after well recovers and during pumping
   Inspect pumping system
   Test chemical and bacterial content of water

Observed: Reduced yield of well.

Possible conditions:
   Plugging of well outside of borehole in the aquifer

Course of action
   Test chemical and bacterial content of water
   Perform videocamera survey of well (to see bacterial growth, scale, and/or cascading water)
   Test pumping system and well
   Measure water levels in remote wells
   Measure static water levels for region

Observed: Reduced yield of well.

Possible conditions:
   Regional drought
   Overuse of the resource
   Recharge area for aquifer is reduced or damaged

Course of action:
   Test pumping system and well
   Measure water levels in remote wells
   Measure static water levels for region
   Test chemical and bacterial content of water
   Perform videocamera survey of well (to see bacterial growth, scale, and/or cascading water)

Observed: Reduced yield of well.

Possible conditions:
   Recharge areas reduced because of seasonal changes
   Permanent (nonseasonal) effects to recharge areas

Course of action:
   Test pumping system and well
   Measure water levels in remote wells
   Measure static water levels for region
   Test chemical and bacterial content of water
   Perform videocamera survey of well (to see bacterial growth, scale, and/or cascading water)
   Review well design

SOLUTIONS TO WELL-MAINTENANCE PROBLEMS

The nature of well deterioration may not be readily discernible during operation. In fact, it may not be recognized until the well actually fails. Since the active portion of a well is located below ground, some degree of neglect may occur. Serious deterioration problems can develop slowly. Once the deterioration processes reach a critical point, they rapidly accelerate to well system failure. If the symptoms of deterioration are recognized before the well reaches this point, rehabilitating the well may be possible. If the well is neglected for too long, the potential for successful rehabilitation is reduced significantly.

The physical characteristics of the newly constructed pumping well should be observed and recorded. This data will become part of the permanent record for the well. The information is used as a performance baseline, and future testing of the pumping well will be compared to the initial values. Initial data to be collected on new wells should include:

1. Drawdown in the well at the design capacity
2. Drawdown in the well at various pumping rates (step test)
3. Specific capacity (calculated from the recorded data)
4. Chemical and biological quality of ground water
5. Sand content of the discharge
6. Plumbness and alignment of the well
7. Wire-to-water efficiency of the pump and pumping system

The permanent records of the well also should include the original well and pump specifications; an as-built construction diagram of the well; the operating characteristics of the installed pump; formation log and mechanical grain-size analysis of aquifer samples and gravel-pack samples (if used). Performance evaluation and the need for rehabilitation action will be determined using this background data.

Routine Measurements, Tests, and Observations

Routine maintenance for water wells requires regular measurements of several performance parameters. The frequency of these measurements is determined on the basis of use. For instance, a well that is used year round should be routinely checked each month or at other convenient intervals. Wells that are operated seasonally should be checked before and after each season. The offseason time can be used to rehabilitate or repair the well or pump. Routine measurements include the following:

- static water level (measured after at least twenty-four-hour recovery)
- yield of the pumping well
- pumping water level at that yield
- comparison of any chemical data

The pumping unit also can be evaluated by means of a quick checklist. The following observations should be made and recorded during routine lubrication and servicing of the pump:

- any increase in sand content in discharge
- excessive heating of motor
- excessive oil consumption
- excessive vibration
- sounds that could possibly indicate cavitation
- cracking or uneven settlement of the pump pad
- settlement or cracking of the ground around the well

The importance of records cannot be overstated. A methodical program of well inspection and monitoring can make it possible to develop a sound diagnosis of the problems. In addition, a regular program of preventive maintenance will ensure that the well provides a reliable source of water.

Measurement data that is recorded under different operating conditions cannot be used in a comparison study. It is extremely important that recorded data be measured under uniform conditions to permit its comparison with historical observations. Only then can a trend be distinguished and a possible problem alleviated.

Data organization and reconciliation are just as important as data collection. Using the data to arrive at real numbers that mean something cannot be emphasized enough. These numbers are the storytellers for well operation. Routine measurements are essential, for they allow the operator to plan and execute a preventive maintenance program before the situation becomes critical. This saves money, time, and effort.

Maintaining Seasonally Operated Wells

Seasonally operated wells, such as irrigation and other large-capacity wells, should begin their maintenance program before the pumping season starts. About two weeks in advance, the static water levels should be measured and recorded. After the start of the pumping season and after each well has been operating continuously for eight hours or more, drawdown, discharge, and sand content of the discharge should be measured. In a multiple well field, each well should be tested individually and the resultant drawdowns in adjacent wells should be observed and recorded during the test.

During the pumping season, static water levels should be measured several times each month. The well is to be shut down for at least twelve hours to get an accurate reading. Discharge and drawdown should be measured and recorded eight or more hours after restart.

At the end of the pumping season, the total depth of each well needs to be determined. If total depth is decreasing over time, this is an indication that sand or other material has accumulated in the bottom of the well. If left unchecked, this sand accumulation will eventually encroach upon the screen and decrease specific capacity. The sand accumulation has to be alleviated in order to prevent further deterioration of the well. The pump should be pulled and inspected for damage and the well should be bailed clean before making any other tests or measurements.

Solving Incrustation Problems

Chemical incrustation problems are treated by using strong acid solutions that chemically dissolve the encrusting inorganic deposits from the well screens, pore spaces of the adjacent aquifer, and other areas that affect well performance. These deposits are most frequently calcium, iron and magnesium carbonates, oxides, hydroxides, and sometimes sulfates.

Physical incrustation is the result of silt- and clay-size particles plugging the well screen and adjacent gravel pack. Sequestering agents, such as phosphates, break the attraction between like ions and enable the fine material to be removed from the well and adjacent aquifer.

Bacterial incrustation problems can be controlled by strong bactericides and oxidizing agents that kill the bacteria and loosen the gelatinous organic material they produce. Combining acid treatments with bactericide treatments is done to dissolve the iron precipitate deposits formed by the bacteria.

Chemicals Used to Restore Water-Well Efficiency

There are three classes of chemicals that are used in water-well maintenance: acids, phosphates, and biocides. Most of the chemicals used in the restoration of water wells may be corrosive or otherwise hazardous. Well restoration should be done by or under the supervision of an experienced person, possibly the representative of a company that specializes in the use of the chemical.

Several applications of the same chemical may be needed to be effective in the removal of encrusting materials. Sometimes different chemicals are used in tandem applications to improve the efficiency of each. The first application of the chemical solution will flow most readily into those areas where the formation is most open, and where resistance to flow is the least. The application will clean these areas and improve permeability. The second application of the chemical will flow in the same pattern as the first unless the solution is vigorously agitated and surged so it is forced into previously unopened areas.

The application techniques and procedures for using a particular chemical are important in the rehabilitation process. Some chemicals are heavier than water, whereas others have the same density or may be even slightly lighter than the water into which the chemical is being applied. Other chemicals react violently when improperly mixed with water. It is important to know about these reactions before rehabilitation begins. It is advisable to consult with an expert or the manufacturer of the chemical before the work is performed.

Once properly introduced into the well, some chemicals require that the well water be stimulated or agitated in some manner, so that the chemical reaction that takes place in the well is most effective. Agitation or well stimulation while treating a well with chemicals can be hazardous. Special procedures must be followed, and again it is wise to consult with experienced people before undertaking a rehabilitation project.

Hazards such as poisonous fumes and gases, and corrosive foam or froth can be especially dangerous to personnel. Agitation is most dangerous when using corrosive additives such as acids or strong oxidizing agents. Experienced individuals and special equipment are necessary. See table 9-1 for a summary of information on well-maintenance chemicals.

Table 9-1. Chemicals Used for Well Maintenance

	Chemical Name	Formula	Application	Concentration
Acids and biocides	Hydrochloric acid	HCl	Carbonate scale, oxides, hydroxides	15%; 2-3 times zone volume
	Sulfamic acid	NH_2SO_3H	Carbonate scale, oxides, hydroxides	15%; 2-3 times zone volume
	Hydroxyacetic acid	$C_2H_4O_3$	Biocide, chelating agent, weak scale removal agent	
	Chlorine	Cl_2	Biocide, sterilization, very weak acid	50-500 ppm
Inhibitors	Diethylthiourea	$(C_2H_5)NCSN(C_2H_5)$	Metal protection	0.2%
	Dow A-73		Metal protection	0.01%
	Hydrated ferric sulfate	$Fe_2(SO_4)_3 \cdot 2\text{-}3H_2O$	For stainless steel	1%
	Aldec 97		With sulfamic acid	2%
	Polyrad 110A		Metal protection	.375%

(continued)

Table 9-1. *(continued)*

	Chemical Name	Formula	Application	Concentration
Chelating agents	Citric acid	$C_6H_8O_7$	Keeps metal ions in solution	
	Phosphate acid	H_3PO_4	Keeps metal ions in solution	
	Rochelle salt	NaOOC(CHOH)$_2$COOK	Keeps metal ions in solution	
	Hydroxyacetic acid	$C_2H_4O_3$	Keeps metal ions in solution	
Wetting agents	Plutonic F-68		Renders a surface nonrepellent to a wetting liquid	
	Plutonic L-62		Renders a surface nonrepellent to a wetting liquid	
Surfactants	Dow F-33		Lowers surface tension of water thereby increasing its cleaning power	
	Sodium tripolyphosphate			
	Sodium hexametaphosphate			

ppm = pounds per minute

10

Consolidated Rock-Well Design

Nowhere is the planning stage and the choice of exploration techniques more important than in hard-rock drilling applications. Drilling in consolidated rock aquifers is not only difficult, but no guarantees exist as to the exact location of water in the formation or the aquifer yields. Almost any degree of investment in the predrilling exploration phase will result in minimizing dry holes and maximizing the efficient use of drilling equipment and personnel.

According to a 1983 National Water Well Association survey, the drilling of a private, domestic well in any type formation costs about $2,000 to $3,000, drilling an irrigation well costs about $8,000, and drilling a large municipal or industrial well costs about $30,000. The importance of reaching an adequate supply of water at a reasonable depth is paramount. Drilling deeply not only costs more during the drilling phase, but also it costs more to produce or pump water from the well.

Drilling in unconsoldiated formations presents fewer significant difficulties as compared with hard-rock drilling. Several advantages exist when using an unconsolidated formation for the source of ground water. Unconsolidated aquifers are shallow and generally uniform in nature, are easily recharged, and generally contain water of better quality than that from consolidated aquifers. In addition, ground water in unconsolidated aquifers moves more uniformly through the aquifer. Long-term yields from wells constructed in unconsolidated formations are easier to predict.

From existing field data (including well logs, and water and formation samples gathered from the general vicinity), the engineer or well driller can better approximate the depth and the yield of a well constructed in an unconsolidated aquifer. Even in areas where available data is lacking, boreholes in unconsolidated formations often can be drilled faster and less expensively (in terms of labor and equipment) than comparable drilling operations that penetrate significant thicknesses of hard rock.

Hard-rock formations—due to their relatively low primary permeability and storage capacities—may yield insignificant amounts of water to a well, even after hundreds of feet of depth have been achieved. If a fracture zone with an adequate yield has been reached, pumping rates from the well may have to be controlled; recharge rates to consolidated rock aquifers may be very limited and are dependent on area rainfall, distance, and depth.

The amount of predrilling exploration funds depends on the type of project, the amount of water that is needed from the well, and the final use of the water that is being developed. Where funding is secondary, investigation and assessment of the ground-water resource in consolidated rock aquifer areas may cover a large geographical area using an array of state-of-the-art exploration methods. If funding is limited or nonexistent, the exploration phase will be scaled down considerably to perhaps an inventory of wells that already exist in the area, their depth and their yields.

Consideration as to the final well system that is best suited to the situation will determine the amount of exploration funding. Large exploration budgets should be expected if the ground-water supply is to be developed from one or two high-production wells. With a limited exploration budget, drilling and developing several low-yielding wells is easier.

FRACTURE CHARACTERISTICS

Consolidated rock aquifers are unique in that the majority of ground water comes from secondary porosity and permeability. The best consolidated rock wells are those that intersect many fractures, joints, and bedding planes, thereby increasing the well's likelihood of producing adequate amounts of water. As a result of secondary porosity and permeability, wells developed in rock aquifers tend to offer limited yields. Consolidated rock formations do not yield water uniformly from the aquifer as do unconsolidated aquifers. Instead, ground water flows from the cracks and fractures into the well at a rate governed by the size, location, and amount of fractures in the well. If few fractures, joints, and bedding planes are present, the yield will be disappointing.

The design of a consolidated rock well is simple and direct. The well must be deep enough to intersect as many fractures and cracks as possible to provide the amount of ground water needed. The well needs no screen or casing once the surface casing is grouted properly into the top of the rock formation.

Finding the optimum location to drill a rock well becomes the highest priority. The right location in the right geologic setting will result in a well that meets or exceeds its demand for ground water. Areas that have experienced tectonic forces in their geologic history are prime candidates for a successful well. Tectonic forces bend and crack the consolidated rock in predictable patterns, creating or increasing the secondary porosity and permeability of the formation. Regional fracture patterns are developed and can be predicted.

Some consolidated rock aquifers are prone to erosion and solution channeling. Limestone and dolomite formations may contain numerous fractures, joints, and bedding planes, as well as many solution features such as underground caves, solution cavities, and sinkholes. While being an important consideration for gaining maximum yields (as for high-production wells), these geologic features must be considered from a pollution standpoint. Solution features do not offer much protection from pollution since some of the highest ground-water flows have been recorded in limestone solution features.

A few consolidated rock aquifers such as sandstone have primary porosity and permeability. The porosity and permeability of these formations is less than that of their counterparts in unconsolidated formations due to the nature of the rock itself. In addition to sorting and packing of individual sand grains, sandstone porosity and permeability is further reduced by the amount and nature of the cement material that binds the grains together, forming the consolidated rock. Calcium carbonate and iron are the two most common cementing materials that bind grains of sand together in sandstone rock.

In some sandstone formations the cement process is mature and the consolidated rock is hard and firm. In other formations the cementing process is young and the sandstone rock is weak and friable, allowing grains of sand to fall into the well or be carried out of the well through the pump with the ground water. In this latter case, the design of the final well should be similar to that of a gravel-packed well in an unconsolidated formation, or sand problems may cause excessive pump repairs and maintenance problems.

EXPLORATION TECHNIQUES FOR CONSOLIDATED ROCK AQUIFERS

To facilitate the discovery of the right location to drill and develop a well, many different exploration techniques can be employed, including the use of elaborate aerial photographic techniques, sophisticated geophysical methods, and simple geologic field techniques. Some of the exploration techniques already have been discussed in chapter 1, so they will be discussed here in general terms.

The first group of exploration techniques include the aerial or reconnaissance methods, which are quite effective in exploring for the optimum location of a consolidated rock well-drilling location. Due to the reliability and low cost of these methods, they are the most widely used for consolidated rock aquifers. Included in this group of techniques are the use of aerial photos and the use of satellite imagery to aid in the location of possible sources of ground water.

Geophysical techniques consist of both surface and borehole methods. Surface geophysical methods include the use of seismic reflection and refraction techniques and electrical resistivity surveys. Borehole methods include the use of caliper logging of drilled boreholes, gamma-ray logging, neutron logging, and other downhole geophysical logs. These downhole logs help determine formation types and water availability.

Geologic field methods consist of a geologist exploring an area, collecting rock samples, measuring exposed rock formations, and predicting where the formation would occur underground at different sites.

Regardless of the method used, the exploration and assessment of ground water in consolidated rock formations should be a phase type of operation. During each of the phases, the designer collects and interprets new pieces of information, making the selection of the next phase of exploration more apparent. Once all of the information is collected or adequate data exists to make a logical and sound decision, the drilling process can begin.

An exploration process should begin with the most obvious and logical phase. All data that is readily available or easily obtained should be collected first. This may include measuring existing wells for depth and yield, reviewing drilling logs for the area, talking with drilling contractors who have drilled in the area, and reviewing all of the written geologic data that exists for the area. Depending on the results, a decision can be made as to whether more extensive techniques should be employed.

Existing Field Data

Existing data should be studied first. Useful data is often obtained from geologic reports, water resource reports, and topographic, geologic, and water resource maps. Geologic reports for an area may contain information on the physical nature of the underlying rock formation, whether it consists of granular material like sandstone or nongranular rocks such as limestone, granites, or basalt.

The use of geologic reports provides a quick overview of the area geology and helps to determine which rock formations will be intercepted by a well and their potential for being an aquifer. Geologic cross sections of an area are useful and can provide such information as rock type, depth and thickness of formations, and changes in the lithology or nature of the rock itself from one area to another. Geologic reports may indicate where outcrops of the formation exist so the designer or geologist can examine the rock firsthand. This is particularly important when sandstones are involved, since the outcrop may give some clue as to the nature and maturity of cementing in the rock, thereby eliminating certain designs.

The reports may discuss the nature and type of tectonic forces to which an area has been subjected. This is important to know if the aquifer rock type is nongranular. Ground-water movement and storage in nongranular rock only occurs through secondary permeability and porosity such as interconnecting fractures, joints, and bedding planes. The more fractures, joints, and bedding planes, the greater the chance of finding appreciable amounts of ground water. Knowing that tectonic forces have existed in an area and knowing the origin of those forces will help predict the type and direction of regional fractures and joints.

There are many types of geologic reports. As a good starting point, some states have an agency responsible for reporting and mapping the geology of their state. Other sources of geologic information can be obtained from the petroleum industry. Although the petroleum industry is more interested in the geology below water-bearing formations, their reports about the nature of tectonic forces, the type and nature of formations, lithologic logs, and cross sections will prove invaluable. Penetrating of shallow formations during the drilling process of a petroleum well will yield cuttings, formation information (whether the shallow formation took drilling fluid or produced water), and other information.

Water resource reports are similar to geologic reports except that the emphasis is placed on the water resources for a particular area. The information typically found in these reports includes an inventory of surface- and ground-water resources. Inventory lists of water wells and their yields are given, as well as stream flow and information about lakes and ponds. Water resource reports will vary in the amount of geologic information and in the quality of the information, but from these reports the geologist often can determine the entire hydrologic scheme of an area and provide a better means of predicting ground water resources.

Topographic maps show elevation changes with a series of parallel lines, each indicating a particular elevation. These changes in the topography may indicate possible ground-water locations. Ground water can typically be found under valleys and along flood plains. These maps may indicate the location and amount of vegetation, springs, wells, and windmills, as well as the presence of surface-water features. Topographic maps are especially useful in hilly areas, karst (limestone) areas, and for locating possible outcrops of rock.

Geologic maps and water resource maps can provide as much useful information as the reports that accompany them. Tectonic maps also will help in the location of major faults, fractures, and joints for an area. Much information can be gained from local consultants and from well drillers and their well logs.

Aerial Photography and Fracture Trace Analysis

High-capacity wells demand a large-scale exploration program. The nature of fractures and faults that are found in consolidated formations such as limestone, dolomite, granite, and other crystalline rocks is indicative of the ability of these rocks to transmit water. These fractures often have subtle surface expressions that reveal their subsurface existence.

The location of high-production wells is often determined by interpreting aerial photographs that allow the hydrogeologist to discern certain surface features. Upon further field investigation and with the aid of aerial photographs, surface fractures, crevices, or joints in the consolidated rock may be located. By finding a location in which two or more fractures, joints, or crevices intersect, a well developed at this location will have a better chance of producing more water. Fractures are usually clustered in zones 5 to 50 feet wide and several thousand feet to more than a mile in length.

Fracture-trace mapping is being used routinely in many areas of the United States to locate wells for individual homes and for municipal and industrial use, attesting to the relatively low cost of the process. Since random drilling in tightly cemented rocks can fail to produce even the minimum amount of ground water to a well, fracture-trace mapping should be applicable in at least part of almost every state. The technique has been successfully applied within a variety of geologic settings in Pennsylvania. Increases in well yields are most spectacular (tenfold or greater) in dense, tightly cemented rock such as marble and limestone. Well yields can be improved in other rock types as shown in table 10-1.

The study of linear features on the earth's surface dates back to the 1800s, but it was not until the early 1960s that much work was done relating photogeologic features to water supply development. This work was preceded by the advent of aerial photography during World War II and the use of aerial photographs by petroleum companies. Petroleum exploration companies found that photolinear features were valuable aids in the exploration of gas and oil.

The terminology of *fracture traces* and *lineaments* was introduced in a paper by Lattman (1958). Lattman and his colleague Parizek (1964) described significantly higher yields from wells that were located on single fracture traces as well as those located on fracture trace

Table 10-1. Application of aerial photograph in locating wells.

Rock Type	Average yield from randomly drilled wells (gpm)	Average yield from aerially located wells (gpm)
Metamorphic	1–10	100–200
Sandstone – tightly cemented	20–30	100–200
Sandstones, siltstones – loosely cemented	100–400	500
Limestone, dolomites, marble	1–20	500–3000

gpm = gallons per minute

intersections. These ideas were based on the analyses of well yields in Pennsylvania and became the basic impetus for much of the current interest in using fracture trace techniques as a way of optimizing well site selection. These techniques have been somewhat restricted by three factors:

1. Lack of appreciation concerning the limitations of this method
2. Lack of precision during drilling and well construction
3. Consideration of other hydrogeologic factors that can influence well yield

The interpretation of aerial photographs depends in part on the characteristics of the images and the classification of the land forms. The image characteristics are determined by tone, pattern, shape, and texture. The difference in tone between various objects, and the sharpness of the boundary between objects or masses, assist in the identification procedure. For example, a dark tone may indicate a basic rock type or a high soil-moisture content. A sharp boundary may indicate the presence of two different rock types or a fracture within a common rock formation.

The classification of the land forms is the next step in the interpretation process. In some areas the geological conditions are closely reflected in morphological expressions of rock types and rock structures, such as fault scarps and lithological boundaries. In these cases, identification can be made with a high degree of certainty. Where surface indications are less evident other factors such as vegetation changes and drainage must be taken into account.

Interpreting aerial photographs requires a critical determination of the exact scale of the photograph. This allows for the accurate transferral of fracture traces from the photograph to the specific ground location. This becomes a very important factor if the zones of fractures that are being searched for are only a few feet wide. The best scale of photograph to use is 1:20,000 (or 1 inch = 1,667 feet), generally viewed as a stereoscopic pair. By using a stereoscope the viewer obtains the mental impression of a three-dimensional model.

To the trained eye, the photo characteristics of hard rocks are not difficult to discern. Armed with the proper information on the area climate (humid, arid, or semiarid) and the knowledge of how different rock types withstand erosion effects, the interpreter is able to ascertain whether certain extrusive rock images are granites, gneisses, quartzites, or other rock types. For example, under humid conditions, granitic rock types generally show a rounded to gently rolling topography.

The linear features that appear in aerial photographs may have various origins. Geological (fractures, faults), vegetational (rows of trees), or man-made (roads, canals) forces all may indicate an apparent linear feature. The fractures can be used to identify certain rock types. For example, fractures in igneous rocks may be irregularly spaced in a crisscross pattern. Widely spaced lineations at right angles to a pronounced topographic trend may indicate metamorphic rock types.

The deeper weathering of fractures and crevices often causes linear valleys that hold deeper soils and higher moisture content. These conditions allow for deeper root development by trees, bushes, and grasses. Plants and trees located along these linear zones grow larger than neighboring plants and trees. In aerial photographs, indications of lush vegetation may appear as alignments of slightly larger tree crowns or branches, slightly thicker bushes, or merely as a darker area due to shadows and higher moisture content. Alternatively, tree rows may only indicate old property lines.

The photogeologic expression of fracture traces and lineaments varies significantly depending on bedrock types and the thicknesses of the overburden soil. For example, fracture trace expressions in soluble limestones can be very obvious, whereas in most other rock types, surface expression of these features is much more subtle. The importance of fracture traces and lineaments to ground-water hydrology is that these surface features are underlain by zones of closely spaced fractures or faults in the bedrock that are capable of transmitting larger quantities of water than surrounding, nonfractured rock.

The fracture-trace technique can be applied with any bedrock type, but the most effective results have occurred in solution limestones that are exposed or have only a thin soil mantle. Some hydrogeologists consider fracture-trace analysis applicable even when a hundred feet or more of glacial cover is present. In such areas of deep overburden, a trained interpreter should be consulted.

It should be noted that fracture-trace analysis does not guarantee water yields or depths where water will be obtained. As effective as the method can be, a small percentage of well locations will have unsatisfactory results. The technique should be considered a means to maximize well yields that are in the upper range possible for a particular geologic and hydrologic setting. Geologists employing this technique will stress this fact, thus avoiding unreasonable expectations. Some geologists and well drillers remain skeptical of the system's efficacy because they have had limited experience with fracture-trace mapping. Much of this skepticism is due to:

- incorrect interpretation of linear features
- imprecise field checking of photogeologic formations
- poor coordination between geologists and well drillers in determining the dip or angle of the fracture, leading to poor results
- a misunderstanding concerning the limitations of the technique

As for ground-water storage characteristics of different rock types, granites and pegmatites are classified as potentially good aquifers. High-grade metamorphics and strongly folded gneisses, especially those having sedimentary origins, are considered poor aquifers.

Test wells should be drilled at the intersection between tensile fractures (which are the main "drains" of the aquifers) and other intersecting fractures. The possible direction of ground-water flow in the fractures can be determined by evaluating topographical conditions. If possible, strikes and dips of surface fractures should be measured and the drilling location calculated based on the surface expression of the features and the angle of dipping. Drilling on the exact location of the intersection may yield disappointing results if the features are dipping at shallow angles.

Satellite Imagery

Satellite imagery is based on the same principle as aerial photography. By using different filtering techniques, the resulting photographs make better use of color infrared film that displays landscape features in different color tones than would appear normally. For example, surface bodies of water may appear black and healthy vegetation may appear red. The use of such photographs distinguishes certain features that will facilitate the search for ground-water sources. Satellite imagery is used to help identify certain rock types such as sandstone or gneiss.

Compared with satellite imagery, aerial photographs have the best spatial resolution, but have the disadvantage of rendering a single recording or static impression of dynamic processes. Through the use of satellite imagery, sequential recordings are possible that permit the observation of seasonal variations such as river discharges or spring flows. Aerial photographs provide details of surface characteristics whereas satellite images show the relationships between regional features and their variations over time.

Geophysical Exploration

In consolidated formations ground water occurs in fractures, fissures, crushed zones, and joints, which may not lend themselves to aerial photography or satellite imagery. Geophysical methods of exploration help locate these fractures by the use of sophisticated electronic instruments. The geophysical properties of water-bearing zones (electrical resistivity, seismic, velocity, average density, and so forth) depend in part on the degree of fracturing, the mode of ground-water occurrence, and the presence of dissolved salts.

In cases where the loose overburden is relatively thick, geophysical methods are used to determine its thickness apart from locating fractures in the bedrock. For example, seismic surveys are used often to determine the thickness of unconsolidated deposits that overlie bedrock. These surveys will help pinpoint the areas where bedrock is thickest, offering the best chance to find coarse unconsolidated sediments or fracture zones. The coarse unconsolidated sediments occur as a result of ancient surface-water bodies and the fracture zones occur near bedrock lows due to their apparent solution and erosion.

Seismic surveys are useful in determining the location of fracture zones and therefore the possibilities for ground water. For instance, a broad zone of intense fracturing tends to take on characteristic electrical and physical properties. Within such a zone, the velocity of seismic waves is much reduced in comparison to velocities in the surrounding hard rock.

Electrical resistivity methods can be used to locate potentially high-yielding fracture zones. Since water-bearing fractures in hard rock are often better electrical conductors than the surrounding rock, these fractures can be detected by electrical mapping. This method is the least difficult in areas where the overburden is uniform and has a fairly high electrical resistivity. However, the overburden is often more electrically complicated, and interpretations will be somewhat more difficult.

DRILLING METHODS IN CONSOLIDATED ROCK

The design and construction of water wells in hard rock is relatively simple. If only a few fractures are expected during the drilling operation, casing is not needed during the drilling process unless the drilling technique itself requires it. Casing in rock wells is only required when drilling through unconsolidated materials on top of the bedrock or when drilling through known cavernous zones in the bedrock itself. Casing is needed, of course, to protect the well and ground-water resource from pollution from surface or near-surface contamination. State well construction codes should be followed when designing any water well.

Dug wells can be excavated in bedrock and can be either drilled or completed by hand. Once the unconsolidated soil zones have been excavated and cased off, the bedrock portion of the well can be dug to its entire depth before casing or a permanent lining is installed. A temporary steel liner can be installed if soft or caving material is encountered.

Percussion and rotary drilling methods are preferred for drilling wells in consolidated rock. If thick layers of unconsolidated material are expected, a combination of drilling methods may be employed. The surface soils are drilled using the conventional mud rotary system, followed by one of the percussion drilling methods, such as air rotary or downhole hammer, for drilling into the bedrock (fig. 10-1). No casing is needed in the consolidated rock portion of the hole, but in wells with overlying unconsolidated materials, casing must be installed in the hole either during the drilling process or right after the unconsolidated soils have been drilled and completed. For complete details on the operation and

10-1. One type of downhole hammer for drilling in bedrock. (Water Well Journal Publishing Company, Columbus, Ohio)

procedures that are used with each of these drilling methods, refer to chapter 3.

Drilling through actual fracture zones may spawn many difficulties that can hinder the drilling progress. Well drillers with little experience with the potentially high yields of fracture zones will purposely move their equipment away from these zones in search of better drilling conditions. With this in mind, certain trade-offs may be necessary to obtain maximum well yields in certain rock formations. Factors such as difficult rotation, loss of circulation (both air and fluid circulation), caving boreholes, and even stuck drill bits are indications of highly fractured zones. Some of these conditions may become so severe that the borehole cannot be completed. However, since adverse drilling factors can indicate high quantities of water, there is good reason to proceed with the drilling operation whenever possible.

Although the drilling of wells in fracture zones increases the likelihood of obtaining higher yields than the average for the area, each well site will not meet with success. After drilling to certain depths in hard rock without evidence of water, it may be wiser to consider spending the time and money on a second well at a second site rather than drilling deeper. A rough guide of 400 to 600 feet is used in the industry as a maximum practical depth for drilling in consolidated formations and locating substantial fracture zones. Below that depth, the weight of the rock and overburden material will close fractures and cracks, reducing the amount of water that will flow into a well.

Recent experiments with magnetic surveys and reports of deep-seated fracture zones occurring in crystalline rock in European countries have led to several studies in the United States to locate deep-seated highly fractured zones of freshwater in the crystalline rock aquifers in New England. Rivendell Resources Inc. of South Hamilton, Massachusetts, and the Riess Foundation, a nonprofit foundation sponsoring this research, have experimented with magnetic surveys to help locate variations that might indicate the presence of ground water occurring under these conditions.

A spokesman for Rivendell states that naturally occurring magnetic lows of the underground rock can be areas in which wells can produce 200 to 300 gallons per minute of water. Wells drilled outside these magnetic lows will yield only 2 to 5 gallons per minute. At the Totten Field Laboratory in Massachusetts, three wells have been drilled varying in depth from 1,500 to 3,000 feet and each well produces more than 400 gallons per minute of water. For the New England area, these wells are considered by many to be gushers.

A maximum depth may be used as guidelines when drilling in deep rock, but by using certain exploration techniques, wells drilled to more than 500 feet—and even more than 1,000 feet—may produce excellent yields. Through a combination of aerial photographs, geophysical methods, good field work, and selection of the proper drilling technique, such depths can become economically feasible in areas where crystalline rock aquifers prevail.

PUMP TESTING ROCK WELLS

The purpose of the pump test is to determine the long-term operating characteristics of the well and the aquifer. The pump test usually involves a hydrologist and the well driller working together. Testing is important because the bail or air lift tests that were run at the time of drilling offer inaccurate estimates of sustainable well yields and little knowledge about the effects on other wells or aquifers in the area when the well is pumped for a long term. Predicting the future behavior of high-yielding fractured-rock wells requires more accurate information. Pump tests are useful in determining the completeness of well development and the quality of the water from the aquifer.

A complete description of running a pumping test is given in chapter 7. The basic equipment necessary includes a pump, a discharge control valve, a flow measuring device, and a hose or pipe to conduct the pumped water away from the well. The pump's capacity is important in that the pump must be large enough to test the well at rates higher than the well's overall pumping rate. This allows for more accurate predictions of future operating levels and regional water level changes.

In addition to the pumping well, a minimum of one observation well or a well already completed in the same aquifer is needed. Whenever feasible, a number of observation wells (both in the fractured rock aquifer and the unconsolidated overburden aquifer) should be drilled for optimum results. The additional observation wells help to assess the geometric complexity of the interconnecting water-bearing fractures.

A common pumping test that is performed on a well completed in a rock aquifer is a step test. A step test involves letting the pump run for short periods of time at rates that begin at low percentages of the anticipated final well yield, and gradually increasing the pumping to a rate far above the final rate. During the so-called steps, the water level in the well is constantly measured, recorded, and plotted on a graph until the drawdown stabilizes. Once stable, the next higher pumping rate can begin. This process is continued until either the rate of pumping is 50 percent over the well's anticipated pumping rate or until the drawdown in the well does not stabilize, indicating that ground-water withdrawal exceeds the aquifer's ability to transmit water.

The resulting graphs have the distinct appearance of steps, hence the name. These graphs become very valuable documents that can be referred to at a later time when another step test is performed on the well. By comparing the original graph with that of the newly created graph, well losses and inefficiencies due to plugging of the aquifer by scale or bacterial growth can be documented. When the well is subsequently rehabilitated, another step test can be performed on the well to determine the effectiveness of the rehabilitation process.

To date, mathematical formulas do not exist for the analysis of a rock well-pumping test. Transmissivity and storativity values are meaningless when obtained from pumping test data. Most of the conditions and assumptions for the unconsolidated sand analysis techniques are not met by a consolidated rock well. The step test is the only valid test and analysis technique currently available.

DEVELOPMENT TECHNIQUES FOR ROCK WELLS

A complete description of well-development techniques is presented in chapter 8. Rock wells are subjected to the same damage to the aquifer that unconsolidated wells are during drilling. Fractures and cracks provide a path of least resistance to mud and air flow and will typically fill up and become clogged with cuttings, which, if not removed, will reduce the overall well yield. Fine particles of sand or cuttings may enter the well during initial pumping of the well, indicating that the well may have clogged fractures.

One basic method of developing a rock well is with a physical brushing of the interior of the borehole with a wire brush. The wire brush is constructed of stiff steel bristles (about 1/8 to 1/4 inch in diameter) extending out from the center of the tool 1 to 2 inches larger than the borehole itself. The contractor lowers the brush into the borehole and by alternately turning and pushing it, begins to dislodge loose material from any clogged fractures. The contractor may brush the well several times before bailing the accumulated material out of the well.

Once brushing is complete, the contractor may opt to surge the well with a surge block or develop the well with compressed air. Either method will cause turbulence in the well and begin the cleanup process. In some wells, caves or vugs in the rock may contain loose sand and gravel material. If the vug or cave is located near the top of the bedrock unit, sand and gravel may have entered the feature from the overlying soil. It may be nearly impossible to remove all of the loose material, and the construction of the well may have to be modified.

Controlled blasting sometimes is used in consolidated rock wells to develop or increase the yield from wells. A blasting expert must be consulted for details, but the technique basically involves the use of explosives set off in a rock well under controlled conditions. The result of this technique is that the well borehole will become highly fractured and possibly intersect neighboring fractures, thus increasing the yield.

Hydrofracturing of wells has been common in the petroleum industry and is used to increase petroleum production in many deep oil wells. It is becoming increasingly popular in the New England areas of the United States. Hydrofracturing involves the injection of water or other fluid into a sealed well, below the casing, under very high pressure. The high-pressure fluid will either create new cracks or crevices in the surrounding bedrock formation or widen existing ones, thus enabling more water to flow freely into the well. The extended cracks and crevices also may intercept larger, regional fractures that transmit additional quantities of ground water.

11

Ground-Water Regions of the United States

CLASSIFICATION OF GROUND-WATER REGIONS

Several divisions of the United States into ground-water regions have been proposed. The most successful and most useful subdivisions were those proposed by Meinzer in 1923 and Thomas in 1952. Meinzer's subdivision scheme was based on the rock unit of the principal aquifer, and consisted of twenty-one ground-water provinces. Thomas combined some of the provinces where differences were minor and his scheme consisted of ten regions. Thomas's classification scheme has been used many times since 1952 to summarize the country's ground-water conditions.

Heath (1984b) subdivided the nation into ground-water regions based on features that affect the availability of ground water. Five such features are incorporated in the Heath scheme:

1. Geologic components of the ground-water system and their arrangement
2. Nature of the water-bearing openings of the dominant aquifer or aquifers with respect to primary and secondary origin
3. Mineral composition of the rock matrix of the dominant aquifers with respect to solubility
4. Water storage and transmission characteristics of the dominant aquifer or aquifers
5. Nature and location of recharge and discharge areas

WESTERN MOUNTAIN RANGES

The Western Mountain Ranges encompass three areas totaling 708,000 square kilometers. The largest area extends in an arc from the Sierra Nevada in California, north through the Coast Ranges and Cascade Mountains in Oregon and Washington, and east and south through the Rocky Mountains in Idaho and Montana, into the Bighorn Mountains in Wyoming and the Wasatch and Uinta Mountains in Utah. The second area includes the southern Rocky Mountains, which extend from the Laramie Range in southeastern Wyoming through central Colorado, into the Sangre de Cristo Range in northern New Mexico. The smallest area includes part of the Black Hills in South Dakota (Heath 1984b).

Geology

The general appearance of the Western Mountain Ranges is one of tall, massive mountains alternating with relatively narrow, steep-sided valleys. Some of the summits and sides of the mountains in much of the region have been carved into distinct shapes by mountain glaciers. These tall mountain ranges can be separated by faulted structural lowlands.

Most of the mountain ranges are underlain by granite and metamorphic rocks. Other ranges, including the San Juan Mountains in southwestern Colorado and the Cascade Mountains in Washington and Oregon, are underlain by lavas and other igneous rocks. The summits and slopes of most of the mountains consist of bedrock exposure or bedrock covered by a layer of boulders and other rock fragments produced by frost action and other weathering processes acting on the bedrock. This layer is generally only a few meters thick on the upper slopes but forms a relatively thick apron along the base of the mountains. The narrow valleys are underlain by relatively thin, coarse, bouldery alluvium washed from the higher slopes. The large valleys are underlain by moderately thick deposits of coarse-grained alluvium transported by streams from the adjacent mountains.

Ground-Water Resources

The Western Mountain Ranges and the mountain ranges in adjacent regions are the principal sources of water supplies developed at lower altitudes in the western half of the United States, because the bulk of precipitation falls here and supplies streams and aquifers by its runoff. Melting snow and precipitation at these high altitudes provide abundant water for ground-water recharge. The thin soils and bedrock fractures fill quickly, and the remaining water runs off overland to streams.

Wells developed in the crystalline bedrock have small yields due to the small storage capacity. These wells are generally adequate only for domestic and stock needs. The best opportunity for ground-water supply development is in the valleys that contain either alluvium or permeable sedimentary material. Large yields are possible in wells developed in these materials.

Problems

The most prevalent well problems in the Western Mountain Ranges region result from silt and clay intrusion into the casings of wells developed in alluvial materials. Scale deposition and biological fouling can be problems. Crystalline well problems are due mainly to fissure plugging by clay and silt intrusion.

ALLUVIAL BASINS

The Alluvial Basins region occupies a discontinuous area of 1,025,000 square kilometers extending from the Puget Sound-Willamette Valley area of Washington and Oregon to western Texas. The region consists of an irregular alternation of basins, valleys, and mountain regions. The difference between a basin and a valley is based on drainage outlets. A basin is a low area surrounded by a topographic divide and has no surface drainage. A valley also is a low area, but has surface drainage to adjacent areas. Although both basins and valleys occur in this region, the word *basin* is used to refer collectively to all areas underlain by alluvial deposits when the distinction is not important. The dominant feature of the Alluvial Basins region is its low relief. The only exception is the Coast Ranges of California, which more closely resemble the Western Mountain Ranges (Heath 1984b).

Geology

The alluvium in the region is derived from erosion of the adjacent mountains, and this material is then transported down by steep gradient streams into the basins where it is deposited in the form of an alluvial fan. Generally, the coarsest material in the alluvial fan is deposited at the apex, adjacent to the mountains, and the material gets progressively finer toward the centers of the basins. These deposits are relatively thick, ranging from several hundred to several thousand meters. The surrounding mountains and bedrock beneath the alluvium deposits consist of granite and metamorphic rocks. These rocks are broken along fractures and faults that may serve as water-bearing openings.

Ground-Water Resources

The alluvial fill functions as an ideal aquifer and creates the opportunity for development of high-yielding wells. At the same time, the Alluvial Basins region is the driest area in the United States. Precipitation ranges from 100 to 400 millimeters per year. However, the mountainous regions receive 400 to 800 millimeters of rain annually. When precipitation does occur in the highlands, it runs off rapidly due to the thin soil layer and infiltrates into the alluvial fans.

The ground water travels through the sand and gravel layers toward the center of the basin. Here ground water may discharge onto the land surface. These discharge areas are called playas. Once the ground water has discharged, it may form a lake during an intense storm, or directly evaporate, leaving a thin deposit of clay and other transported sediments and a crust of insoluble salts that were dissolved in the water.

Problems

The most prevalent well problems in the Alluvial Basins region are clay, silt and sand intrusion, scale deposition, high iron content, biological fouling, casing failure, and limited recharge. A related problem is the result of heavy dependence on ground water. With continued withdrawals of more water than the amount replenished by recharge, local and regional subsidence of the land surface is the result.

The valley alluvium deposits consist of sand and gravel layers that are interbedded with finer grained layers of clay and silt. When the hydraulic head or water table is lowered by withdrawals, the saturated silt and clay layers move slowly into the sand and gravel layers. This movement may take several years to become noticeable. Already areas in southern Arizona have observed land subsidence of more than 4 meters.

COLUMBIA LAVA PLATEAU

The Columbia Lava Plateau occupies an area of 366,000 square kilometers in northeastern California, eastern Washington and Oregon, southern Idaho, and northern Nevada. Standing at an altitude generally between 500 and 1,800 meters above sea level, the plateau is underlain by a great thickness of lava flows irregularly interbedded with silt, sand, and other unconsolidated deposits. The plateau is bordered on the west by the Cascade Range, on the north by the Okanogan Highlands, and on the east by the Rocky Mountains. On the south it grades into the Alluvial Basins region, as the area occupied by lava flows decreases and the typical basin and range topography of the Alluvial Basins region gradually prevails. Altitudes in a few of the mountainous areas exceed 3,000 meters.

Geology

The Columbia Lava Plateau is composed of a thick sequence of lava flows that are irregularly interbedded with thin unconsolidated deposits and overlain by thin soils. The lava flows, which range in thickness from less than 50 meters adjacent to the bordering mountain ranges to more than 1,000 meters in south-central Washington and southern

Idaho, are the principal water-bearing unit in the region. The water-bearing lava is underlain by granite, metamorphic rocks, older lava flows, and sedimentary rocks, none of which are very permeable. Individual lava flows in the water-bearing zone range in thickness from several meters to more than 50 meters and average about 15 meters.

Ground-Water Resources

The volcanic rocks yield water mainly from permeable zones that occur at or near the contacts between some flow layers. The origin of these flow-contact or interflow zones is complex and involves the relatively rapid cooling of the top of flows, resulting in formation of a crust. As the molten lava beneath continues to flow, the crust may be broken into a rubble of angular fragments that in places contains numerous holes where gas bubbles formed, giving the rock the appearance of a frozen froth. The slower cooling of the central and lower parts of the thicker flows results in a dense, flintlike rock that in the lower part contains widely spaced, irregular fractures. The flows grade upward into a zone containing closely spaced vertical fractures that break the rock into a series of hexagonal columns (Newcomb 1961).

From the standpoint of the hydraulic characteristics of the volcanic rocks, it is useful to divide the Columbia Lava Plateau region into two parts: (1) the area in southeastern Washington, northeastern Oregon, and the Lewiston area of Idaho, part of which is underlain by volcanic rocks of the Columbia River Group; and (2) the remainder of the area including the Snake River Plain. The basalt underlying the Snake River Plain is referred to as the Snake River Basalt; the basalt underlying southeastern Oregon and the remainder of this area has been divided into several units to which names of local origin are applied (Hampton 1964).

The Columbia River Group consists of thick flows that have been deformed into a series of broad folds and offset locally along normal faults. Movement of ground water occurs primarily through the interflow zones near the top of flows and, to a much smaller extent, through fault zones and through joints developed in the dense central and lower parts of the flows. The axes of sharp folds and the offset of the interflow zones along faults form subsurface dams that affect the movement of ground water. Water reaching the interflow zones tends to move down the dip of the flows from fold axes and collects updip behind faults that are transverse to the direction of movement (Newcomb 1961). As a result, the basalt in parts of the area is divided into a series of barrier-controlled reservoirs that are poorly connected hydraulically to adjacent reservoirs.

The water-bearing basalt underlying California, Nevada, southeastern Oregon, and southern Idaho consists of small, thin flows that have been affected to a much smaller extent by folding and faulting than has the Columbia River Group. The thin flows contain extensive, highly permeable interflow zones that are effectively interconnected through a dense network of cooling fractures. Structural barriers to ground-water movement, such as those of the Columbia River Group, are of minor importance.

Much of the Columbia Lava Plateau region is in the rain shadow east of the Cascades and, as a result, receives only 200 to 1,200 millimeters of precipitation annually. Recharge to the ground-water system depends on the amount and seasonal distribution of precipitation and the permeability of the surficial materials. Most precipitation occurs in the winter and thus coincides with the cooler, nongrowing season when conditions are most favorable for recharge.

The large withdrawal of water in the Columbia Lava Plateau for irrigation, industry, and other uses has resulted in declines in groundwater levels of as much as 30 to 60 meters in several areas. In most of these areas the declines have been slowed or stopped through regulatory restrictions or other changes that have reduced withdrawals. Declines are still occurring in some areas at rates of a few meters per year.

COLORADO PLATEAU AND WYOMING BASIN

The Colorado Plateau and Wyoming Basin region occupies an area of 414,000 kilometers in Arizona, Colorado, New Mexico, Utah, and Wyoming. It is a region of canyons and cliffs; of thin, patchy, rocky soils; and of sparse vegetation adapted to the arid and semiarid climate. The large-scale structure of the region is that of a broad plateau standing at an altitude of 2,500 to 3,500 meters and underlain by essentially horizontal to gently dipping layers of consolidated sedimentary rocks. The plateau structure has been modified by an irregular alternation of basins and domes, in some of which major faults have caused significant offset of the rock layers.

The region is bordered on the east, north, and west by mountain ranges that tend to obscure its plateau structure. The northern part of the region—the part occupied by the Wyoming Basin—borders the Nonglaciated Central region at the break in the Rocky Mountains between the Laramie Range and the Bighorn Mountains. The region contains small, isolated mountain ranges, and extinct volcanoes and lava fields are widely scattered over the region.

Geology

The Colorado Plateau and Wyoming Basin region consists mainly of thin soils overlaying consolidated rocks. The rocks that underlie the region consist principally of sandstone shale, and limestone of Paleozoic to Cenozoic age. In parts of the region these rock units include significant amounts of gypsum (calcium sulfate). In the Paradox Basin in western Colorado the rock units include thick deposits of sodium- and potassium-bearing minerals, principally halite (sodium chloride). The sandstones and shales are most prevalent and most extensive in occurrence. The sandstones are the principal sources of ground water in the region and contain water in fractures developed along bedding planes, across the beds, and in interconnected pores. The most productive sandstones are those in which calcium carbonate or other cementing material has been deposited only around the point of contact of the sand grains. Thus, many of the sandstones are only partially cemented and retain significant primary porosity.

Unconsolidated deposits are of relatively minor importance in this region. Thin deposits of alluvium capable of yielding small to moderate supplies of ground water occur along parts of the valleys of major streams, especially adjacent to the mountain ranges in the northern and eastern parts of the region. These deposits are partly of glacial origin. In most of the remainder of the region, there are large expanses of exposed bedrock, and the soils, where present, are thin and rocky.

Erosion has produced extensive lines of prominent cliffs in the region. The tops of these cliffs are generally underlain and protected by resistant sandstones. Erosion of the domes has produced a series of concentric, steeply dipping ridges, also developed on the more resistant sandstones.

Ground-Water Resources

Recharge of the sandstone aquifers occurs where they are exposed above the cliffs and in the ridges. Average precipitation ranges from about 150 millimeters in the lower areas to about 1,000 millimeters in the higher mountains. The heaviest rainfall occurs in the summer in isolated, intense thunderstorms during which some recharge occurs where intermittent streams flow across sandstone outcrops. However, most recharge occurs in the winter during snowmelt periods. Water moves down the dip of the beds away from the recharge areas to discharge along the channels of major streams through seeps and springs and along the walls of canyons cut by the streams.

The quantity of water available for recharge is small, but so are the porosity and the transmissivity of most of the sandstone aquifers. Because of the absence of a thick cover of unconsolidated rock in the recharge areas, there is little opportunity for such materials to serve as a storage reservoir for the underlying bedrock. The water in the sandstone aquifers is unconfined in the recharge areas and is confined downdip. Because most of the sandstones are consolidated, the storage coefficient in the confined parts of the aquifers is very small. This small storage coefficient, together with the small transmissivities, results in even small rates of withdrawal, causing extensive cones of depression around pumping wells.

Springs exist at places near the base of the sandstone aquifers where they crop out along the sides of canyons. Discharge from the springs results in dewatering the upper parts of the aquifer some distance back from the canyon walls.

Problems

The Colorado Plateau and Wyoming Basin is a dry, sparsely populated region in which most water supplies are obtained from the perennial streams that flow across it from the bordering mountains. Less than 5 percent of the water needs are supplied by ground water, and the development of even small ground-water supplies requires the application of considerable knowledge of the occurrence of both rock units and their structure, and of the chemical quality of the water. Because of the large surface relief and the dip of the aquifers, wells even for domestic or small livestock supplies must penetrate to depths a few hundred meters in much of the area. Thus the development of ground-water supplies is far more expensive than in most other parts of the country. However, ground water in the region can support a substantial increase over the present withdrawals.

As in most other areas of the country underlain by consolidated sedimentary rocks, mineralized (saline) water—water containing more than 1,000 mg/l of dissolved solids—is widespread. Most of the shales and siltstones contain mineralized water throughout the region and below altitudes of about 2,000 meters. Freshwater—water containing less than 1,000 mg/l of dissolved solids—occurs only in the most permeable sandstones and limestones. Much of the mineralized water is due to the solution of gypsum and halite by water circulating through beds that contain these minerals. Although the aquifers that contain mineralized water are overlain by aquifers that contain freshwater, the situation is reversed in a few places where aquifers containing mineralized water are underlain by more permeable aquifers containing freshwater.

HIGH PLAINS

The High Plains region occupies an area of 450,000 square kilometers extending from South Dakota to Texas. The Plains are a remnant of a great alluvial plain built in Miocene time by streams that flowed east from the Rocky Mountains. The original depositional surface of the alluvial plain is still almost unmodified in large areas, especially in Texas and New Mexico, and forms a flat, imperceptibly eastward-sloping tableland that ranges in altitude from about 2,000 meters near the Rocky Mountains to about 500 meters along its eastern edge. The surface of the southern High Plains contains numerous playas, which hold water following heavy rains. Some geologists believe these shallow, circular depressions are due to solution of soluble materials by percolating water and accompanying compaction to the alluvium.

Geology

The High Plains region is underlain by one of the most productive and most intensively developed aquifers in the United States. The alluvial materials derived from the Rocky Mountains, which are referred to as the Ogallala Formation, are the dominant geologic unit of the High Plains aquifer. The Ogallala ranges in thickness from a few meters to more than 200 meters and consists of poorly sorted and generally unconsolidated clay, silt, sand, and gravel.

Older geologic units that are hydrologically connected to the Ogallala and thus form a part of the High Plains aquifer include the Arikaree Group of Miocene age and a small part of the underlying Brule Formation. The Arikaree Group underlies the Ogallala in parts of western Nebraska, southwestern South Dakota, southeastern Wyoming, and northern Colorado. It is predominantly a massive, very fine- to fine-grained sandstone that locally contains beds of volcanic ash, silty sand, and sandy clay. The maximum thickness of the Arikaree is about 300 meters, in western Nebraska. The Brule Formation of Oligocene age underlies the Arikaree. In most of the area in which it occurs, the Brule forms the base of the High Plains aquifer.

Ground-Water Resources

Prior to the erosion that removed most of the western part of the Ogallala, the High Plains aquifer was recharged by the streams that flowed onto the plain from the mountains to the west as well as by local precipitation. The only source of recharge now is local precipitation, which ranges from about 400 millimeters along the western boundary of the region to about 600 millimeters along the eastern boundary. Precipitation and ground-water recharge on the High Plains vary in an east-west direction, but recharge to the High Plains aquifer also varies in a north-south direction. The average annual rate of recharge has been determined to range from about 5 millimeters in Texas and New Mexico to about 100 millimeters in the Sand Hills in Nebraska. This large difference is explained by differences in evaporation and transpiration, and by differences in the permeability of the surficial materials.

In some parts of the High Plains, especially in the southern part, the near-surface layers of the Ogallala form a material of relatively low permeability called caliche. Precipitation on areas underlain by caliche soaks slowly into the ground. Much of this precipitation collects in playas that are underlain by silt and clay, which hamper infiltration, with the result that most of the water is lost to evaporation. During years of average or below-average precipitation, all or nearly all of the precipitation is returned to the atmosphere by evapotranspiration. Thus, it is only during years of excessive precipitation that significant recharge occurs and this, as noted previously, averages about only 5 millimeters per year in the southern part of the High Plains. In the Sand Hills area, the lower evaporation and transpiration and the permeable sandy soil results in about 20 percent of the precipitation reaching the water table as recharge.

Natural discharge from the aquifer occurs to streams, springs, and seeps along the eastern boundary of the plains and by evaporation and transpiration in areas where the water table is within a few meters of the land surface. However, at present the largest discharge is probably through wells. The widespread occurrence of permeable layers of sand and gravel, which permit the construction of large-yield wells almost anyplace in the region, has led to the development of an extensive agricultural economy largely dependent on irrigation.

Problems

Most irrigation water is derived from ground-water storage, resulting in a long-term decline in ground-water levels of as much as 1 meter per year in parts of the region. The lowering of the water table has resulted in a 10 to 50 percent reduction in the saturated thickness of the High Plains aquifer in an area of 130,000 square kilometers. The largest reductions have occurred in the Texas panhandle and in parts of Kansas and New Mexico.

The depletion of ground-water storage in the High Plains, as reflected in the decline in the water table and the reduction in the saturated thickness, is a matter of increasing concern. From the standpoint of the region as a whole, the depletion does not yet represent a large part of the storage that is available for use, but in areas where intense irrigation has long been practiced, depletion of storage is severe.

NONGLACIATED CENTRAL REGION

The Nonglaciated Central region is an area of about 1,737,000 square kilometers extending from the Appalachian Mountains on the east to the Rocky Mountains on the west. The part of the region in eastern Colorado and northeastern New Mexico is separated from the remainder of the region by the High Plains region. The Nonglaciated Central region includes the Triassic Basins in Virginia and North Carolina and the "driftless" area in Wisconsin, Minnesota, Iowa, and Illinois, where glacial deposits, if present, are thin and of no hydrologic importance. The region is a topographically complex area that ranges from the Valley and Ridge section of the Appalachian Mountains on the east, westward across the Great Plains to the foot of the Rocky Mountains. It includes, among other hilly or mountainous areas, the Ozark Plateaus in Missouri and Arkansas. Altitudes range from 150 meters above sea level in central Tennessee and Kentucky to 1,500 meters along the western boundary of the region.

Geology

The Nonglaciated Central region is geologically complex. Most of it is underlain by consolidated sedimentary rocks that range in age from Paleozoic to Tertiary and consist largely of sandstone, shale, carbonate rocks (limestone and dolomite), and conglomerate. A small area in Texas and western Oklahoma is underlain by gypsum. Throughout most of the region the rock layers are horizontal or gently dipping.

The land surface in most of the region is underlain by regolith formed by chemical and mechanical breakdown of the bedrock. In the western part of the Great Plains the residual soils are overlain by or intermixed with eolian (wind-laid) deposits. The thickness and composition of the regolith depend on the composition and structure of the parent rock and on the climate, land cover, and topography. In areas underlain by relatively pure limestone the regolith consists mostly of clay and is only a few meters thick. Where the limestones contain chert and in areas underlain by shale and sandstone, the regolith is thicker, up to 30 meters or more in some areas. The chert and sand form moderately permeable soils, whereas the soils developed on shale are finer-grained and less permeable.

Ground-Water Resources

The principal water-bearing openings in the bedrock are fractures along which the rocks have been broken by stresses imposed on the earth's crust at different times since the rocks were consolidated. The fractures occur in three sets. The first set, and the one that is probably of greatest importance from the standpoint of ground water and well yields, consists of fractures developed along bedding planes. Where the sedimentary layers making up the bedrock are essentially horizontal, the bedding-plane fractures are more or less parallel to the land surface. The two remaining sets of fractures are essentially vertical and thus cross the bedding planes at a steep angle. The primary difference between the sets of vertical fractures is in the orientation of the fractures in each set. For example, in parts of the region one set of vertical fractures is oriented in a northwest-southeast direction and the other set in a northeast-southwest direction. The vertical fractures facilitate movement of water across the rock layers and thus serve as the principal hydraulic connection between the bedding-plane fractures.

In the parts of the region in which the bedrock has been folded or bent, the occurrence and orientation of fractures are more complex. In these areas the dip of the rock layers and the associated bedding-plane fractures range from horizontal to vertical. Fractures parallel to the land surface, where present, are probably less numerous and of more limited extent than in areas of flat-lying rocks.

Recharge of the ground-water system in this region occurs primarily in the outcrop areas of the bedrock aquifers in the uplands between streams. Precipitation in the region ranges from about 400 millimeters per year in the western part to more than 1,200 millimeters in the eastern part. This wide difference in precipitation is reflected in recharge rates, which range from about 5 millimeters per year in west Texas and New Mexico to as much as 500 millimeters per year in Pennsylvania and eastern Tennessee. Discharge for the ground-water system is by springs and seepage into streams, and by evaporation and transpiration in areas where the water table is within a few meters of land surface.

Problems

Yields of wells in most of the region are small, making the Nonglaciated Central region one of the least favorable ground-water regions in the country. Even in parts of the areas underlain by cavernous limestone, yields are moderately low because of both the absence of a thick regolith and the large water-transmitting capacity of the cavernous openings, which quickly discharge the water that reaches them during periods of recharge.

The exceptions to the small well yields are the cavernous limestones of the Edwards Plateau, the Ozark Plateaus, and the Ridge and Valley section. The Edwards Plateau in Texas is bounded on the south by the Balcones Fault Zone, in which limestone and dolomite up to 150 meters in thickness have been extensively faulted. The faulting has facilitated the development of solution opening, making this zone one of the most productive aquifers in the country.

Another feature that makes much of the Nonglaciated Central region unfavorable for ground-water development is the occurrence of salty water at relatively shallow depths. Most of the salty water is believed to be connate, that is, it was trapped in the rocks when they emerged from the sea in which they were deposited. Other possible sources include: (1) seawater that entered the rocks during a later time when the land again was beneath the sea and (2) salty water derived from solution of salt beds that underlie parts of the region.

The presence of connate water at shallow depths is doubtless due to several factors, including, in the western part of the area, a semiarid climate and, consequently, a small rate of recharge. Other factors probably include an extremely slow rate of ground-water circulation at depths greater than a few hundred meters.

GLACIATED CENTRAL REGION

The Glaciated Central region occupies an area of 1,297,000 square kilometers extending from the Triassic Basin in Connecticut and Massachusetts and the Catskill Mountains in New York on the east to the northern part of the Great Plains in Montana on the west. The part of the region in New York and Pennsylvania is characterized by rolling hills and low, rounded mountains that reach altitudes of 1,500 meters. Westward across Ohio to the western boundary of the region along the Missouri River, the region is flat to gently rolling. Among the more prominent topographic features in this part of the region are low, relatively continuous ridges (moraines), which were formed at the margins of ice sheets that moved southward across the area one or more times during the Pleistocene.

Geology

The Glaciated Central region is underlain by flat-lying consolidated sedimentary rocks that range in age from Paleozoic to Tertiary. They consist primarily of sandstone, shale, limestone, and dolomite. The bedrock is overlain by glacial deposits which, in most of the area, consist chiefly of till, an unsorted mixture of rock particles deposited directly by the ice sheets. The till is interbedded with and overlain by sand and gravel deposited by meltwater streams, by silt and clay deposited in glacial lakes, and, in large parts of the north-central states, by loess, a well-sorted silt believed to have been deposited primarily by the wind.

Large parts of the region are underlain by limestones and dolomites in which the fractures have been enlarged by solution. Caves are common in the limestones where the ice sheets were relatively thin, as near the southern boundary of the region and in the driftless area.

Ground-Water Resources

Ground water occurs both in the glacial deposits and in the bedrock. Water occurs in the glacial deposits in pores between the rock particles and in the bedrock primarily along fractures. The dominant water-bearing fractures in the bedrock are along bedding planes. Water also occurs in the bedrock in steeply dipping fractures that cut across the bedsand, in some sandstone conglomerates, and in primary pores that were not destroyed in the process of cementation and consolidation.

The glacial deposits are recharged by precipitation on the interstream areas, and serve as a source of water to shallow wells and as a reservoir for recharge to the underlying bedrock. Ground water in small to moderate amounts can be obtained anyplace in the region, both from the glacial deposits and from the bedrock. Large to very large amounts are obtained from the sand and gravel deposits and from some of the limestones, dolomites, and sandstones in the north-central states. The shales are the least productive bedrock formations in the region.

Problems

Because of the widespread occurrence of limestone and dolomite, water from both the glacial deposits and the bedrock contains as much as several hundred milligrams per liter of dissolved minerals and is moderately hard. Concentrations of iron in excess of 0.3 mg/l is a problem in water from some of the sandstone aquifers in Wisconsin and Illinois, and locally in glacial deposits throughout the region. Sulfate in excess of 250 mg/l is a problem in water both from the glacial deposits and from the bedrock in parts of New York, Ohio, Indiana, and Michigan.

As is the case in the Nonglaciated Central region, mineralized water occurs at shallow depth in the bedrock in large parts of the Glaciated Central region. Because the principal constituent in the mineralized water is sodium chloride, the water is commonly referred to as saline or salty. The thickness of the freshwater zone in the bedrock depends on the vertical hydraulic conductivity of both the bedrock and the glacial deposits, and on the effectiveness of the hydraulic connection between them. Both the freshwater and the underlying saline water move toward the valleys of perennial streams to discharge. As a result, the depth to saline water is less under valleys than under uplands, because of lower altitudes and because of the upward movement of the saline water to discharge. In those parts of the region underlain by saline water, the concentration of dissolved solids increases with depth. At depths of 500 to 1,000 meters in much of the region, the mineral content of the water approaches that of seawater. At greater depths, the mineral content may reach concentrations several times that of seawater.

PIEDMONT AND BLUE RIDGE REGION

The Piedmont and Blue Ridge region is an area of about 247,000 square kilometers extending from Alabama on the south to Pennsylvania on the north. The Piedmont part of the region consists of low, rounded hills and long, rolling, northeast-southwest trending ridges, whose summits range from about a hundred meters above sea level along its eastern boundary with the Coastal Plain, to 500 to 600 meters along its boundary with the Blue Ridge area to the west. The Blue Ridge is mountainous and includes the highest peaks east of the Mississippi. The mountains, some of which reach altitudes of more than 2,000 meters, have smooth rounded outlines and are bordered by well-graded streams flowing in narrow valleys.

Geology

The Piedmont and Blue Ridge region is underlain by bedrock of Precambrian and Paleozoic age consisting of igneous and metamorphosed igneous and sedimentary rocks. These include granite, gneiss, schist, quartzite, slate, marble, and phyllite. The land surface in the Piedmont and Blue Ridge is underlain by clay-rich, unconsolidated material derived from in-situ weathering of the underlying bedrock. This material, which averages about 10 to 20 meters in thickness and may be as much as 100 meters thick on some ridges, is referred to as saprolite. In many valleys, especially those of larger streams, flood plains are underlain by thin, moderately well-sorted alluvium deposited by the streams. When the distinction between saprolite and alluvium is not important, the term *regolith* is used to refer to the layer of unconsolidated deposits.

Ground-Water Resources

The regolith contains water in pore spaces between rock particles. The bedrock does not have any significant intergranular porosity. It contains water, instead, in sheetlike openings formed along fractures (that is, breaks in the otherwise solid rock). The hydraulic conductivities of the regolith and the bedrock are similar and range from about 0.001 to 1 m day^{-1}. The major difference in their water-bearing characteristics is their porosities, that of regolith being about 20 to 30 percent and that of the bedrock about 0.01 to 2 percent. Small supplies of water adequate for domestic needs can be obtained from the regolith through large-diameter bored or dug wells. However, most wells, especially those where moderate supplies of water are needed, are small in diameter and are cased through the regolith and finished with open holes in the bedrock. Although as noted, the hydraulic conductivity of the bedrock is similar to that of the regolith, bedrock wells have much larger yields than regolith wells because, being deeper, they have a much larger available drawdown.

All ground-water systems function both as reservoirs that store water and as pipelines (or conduits) that transmit water from recharge areas to discharge areas. The yield of bedrock wells in the Piedmont and Blue Ridge region depends on the number and size of fractures penetrated by the open hole and the replenishment of the fractures by seepage into them from the overlying regolith. Thus, the ground-water system in this region can be viewed, from the standpoint of ground-water development, as a terrane in which the reservoir and pipeline functions are effectively separated. Because of its larger porosity, the regolith functions as a reservoir that slowly feeds water downward into the fractures in the bedrock. The fractures serve as an intricate interconnected network of pipelines that transmit water either to springs or streams, or to wells.

Because the yield of bedrock wells depends on the number of fractures penetrated by the wells, the key element in selecting well sites is recognizing the relation between the present surface topography and the location of fractures in the bedrock. Most of the valleys, draws, and other surface depressions indicate the presence of more intensely fractured zones in the bedrock, which are more susceptible to weathering and erosion than are the intervening areas. Because fractures in the bedrock are the principal avenues along which ground water moves, the best well sites appear to be in draws on the sides of the valleys of perennial streams where the bordering ridges are underlain by substantial thicknesses of regolith. Wells located at such sites seem to be most effective in penetrating open water-bearing fractures and in intercepting ground water draining from the regolith. Chances of success seem to be somewhat less for wells on the floodplains of perennial streams, possibly because the alluvium obscures the topographic expression of bedrock fractures. The poorest sites for wells are on the tops of ridges and mountains where the regolith cover is thin or absent and the bedrock is sparsely fractured.

As a general rule, fractures near the bedrock surface are most numerous and have the largest openings, so that the yield of most wells is not increased by drilling to depths greater than about 100 meters. Exceptions to this occur in Georgia and some other areas where water-bearing, low-angle faults or fractured zones are present at depths as great as 200 to 300 meters.

Problems

The Piedmont and Blue Ridge region has long been known as an area unfavorable for ground-water development. This reputation seems to have resulted both from the small reported yields of the numerous domestic wells in use and from a failure to apply existing technology to the careful selection of well sites where moderate yields are needed. As water needs in the region increase and as reservoir sites on streams become increasingly more difficult to obtain, it will be necessary to make more intensive use of ground water.

NORTHEAST AND SUPERIOR UPLANDS

The Northeast and Superior Uplands region is made up of two separate areas totalling about 415,000 square kilometers. The Northeast Upland encompasses the Adirondack Mountains, the Lake Champlain valley, and nearly all of New England. The parts of New England not included are the Cape Cod area and nearby islands, which are included in the Glaciated Central region. The Superior Upland encompasses most of the northern parts of Minnesota and Wisconsin adjacent to the western end of Lake Superior. The region is characterized by rolling hills and low mountains.

Geology

Bedrock in the region ranges in age from Precambrian to Paleozoic and consists mostly of granite, syenite, anorthosite, and other intrusive igneous rocks and metamorphosed sedimentary rocks consisting of gneiss, schist, quartzite, slate, and marble. Most of the igneous and metamorphosed sedimentary rocks have been intensely folded and cut by numerous faults.

The bedrock is overlain by unconsolidated deposits laid down by ice sheets that covered the areas one or more times during the Pleistocene and by gravel, sand, silt, and clay laid down by meltwater streams and in lakes that formed during the melting of the ice. The thickness of the glacial deposits ranges from a few meters on the higher mountains, which also have large expanses of barren rock, to more than 100 meters in some valleys. The most extensive glacial deposit is till, which was laid down as a nearly continuous blanket by the ice, both in valleys and on the uplands. In most of the valleys and other low areas, the till is covered by glacial outwash consisting of interlayered sand and gravel, ranging in thickness from a few meters to more than 20 meters, that was deposited by streams supplied by glacial meltwater. In several areas, including parts of the Champlain Valley and the lowlands adjacent to Lake Superior, the unconsolidated deposits consist of clay and silt deposited in lakes that formed during the melting of the ice sheets.

Ground-Water Resources

Ground-water supplies are obtained in the region from both the glacial deposits and the underlying bedrock. The largest yields come from the sand and gravel deposits, which in parts of the valleys of large streams are as much as 60 meters thick. Water occurs in the bedrock in fractures similar in origin, occurrence, and hydraulic characteristics to those in the Piedmont and Blue Ridge region. In fact, the primary difference in ground-water conditions between the Piedmont and Blue Ridge region and the Northeast and Superior Uplands region is related to the materials that overlie the bedrock. In the Piedmont and Blue Ridge, these consist of unconsolidated material derived from weathering of the underlying bedrock. In the Northeast and Superior Uplands, the overlying materials consist of glacial deposits which, having been transported either by ice or by streams, do not have a composition and structure controlled by that of the underlying bedrock. These differences in origin of the regolith between the two regions are an important consideration in the development of water supplies.

The glacial deposits in the region serve as a storage reservoir for the fractures in the underlying bedrock, in the same way the saprolite functions in the Piedmont and Blue Ridge region. The major difference is that the glacial deposits on hills and other upland areas are much thinner than the saprolite in similar areas in the Piedmont and Blue Ridge and, therefore, have a much smaller ground-water storage capacity.

Water supplies in the Northeast and Superior Uplands region are obtained from open-hole drilled wells in sand and gravel, and from large-diameter bored or dug bedrock, especially in the Superior Upland. Water supplies are more uncertain in this region than from the fractured rocks in the Piedmont and Blue Ridge because the ice sheets that advanced across the Northeast and Superior Uplands removed the upper, more fractured part of the rock and also tended to obscure many of the fracture-caused depressions in the rock surface with the layer of glacial till. Thus, use of surface depressions in the Northeast and Superior Uplands region to select sites of bedrock wells is not as satisfactory as in the Piedmont and Blue Ridge.

Problems

Most of the rocks that underlie the Northeast and Superior Uplands are relatively insoluble, and, consequently, the ground water in both the glacial deposits and the bedrock contains less than 500 mg/l of dissolved solids. Two of the most significant water-quality problems confronting the region, especially the Northeast Upland section, are acid precipitation and pollution caused by salts used to de-ice highways. Much of the precipitation falling on the Northeast in 1982 had a pH in the range of 4 to 6 units. Because of the low buffering capacity of the soils derived from the rocks underlying the area, there is little opportunity for the pH to be increased. One of the results of this is the gradual elimination of living organisms from many lakes and streams. The effect on ground-water quality, which will develop much more slowly, has not been determined. De-icing salts affect ground water adjacent to streets and roads maintained for winter travel.

ATLANTIC AND GULF COASTAL PLAIN

The Atlantic and Gulf Coastal Plain region is an area of about 844,000 square kilometers extending from Cape Cod, Massachusetts on the north to the Rio Grande in Texas on the south. This region ranges in width from a few kilometers near its northern end to nearly a thousand kilometers in the vicinity of the Mississippi River. The great width near the Mississippi reflects the effect of a major downwarped zone in the earth's crust that extends from the Gulf of Mexico to about the confluence of the Mississippi and Ohio Rivers.

Geology

The region is underlain by unconsolidated sediments that consist principally of sand, silt, and clay transported by streams from the adjoining uplands. These sediments, which range in age from Jurassic to the present, range in thickness from less than a meter near the inner edge of the region to more than 12,000 meters in southern Louisiana. The greatest thicknesses are along the seaward edge of the region and along the axis of the Mississippi embayment. The sediments were deposited on floodplains and as deltas where streams reached the coast and, during different invasions of the region by the sea, were reworked by waves and ocean currents. Thus, the sediments are complexly interbedded to the extent that most of the named geologic units into which they have been divided contain layers of the different types of sediment that underlie the region. These named geologic units (or formations) dip toward the coast or toward the axis of the Mississippi embayment, with the result that those that crop out at the surface form a series of bands roughly parallel to the coast or to the axis of the embayment. The oldest formations crop out along the inner margin of the region, and the youngest crop out in the coastal area. Within any formation the coarsest-grained materials (sand, at places interbedded with thin gravel layers) tend to be most abundant near source areas. Clay and silt layers become thicker and more numerous downdip.

Ground-Water Resources

From the standpoint of well yields and ground-water use, the Atlantic and Gulf Coastal Plain is one of the most important regions in the country. Recharge to the ground-water system occurs in the interstream areas, both where sand layers crop out and by percolation downward across the interbedded clay and silt layers. Discharge from the system occurs by seepage to streams, estuaries, and the ocean.

Wells that yield moderate to large quantities of water can be constructed almost anywhere in the region. Because most of the aquifers consist of unconsolidated sand, wells require screens; where the sand is fine-grained and well sorted, the common practice is to surround the screens with a coarse sand or gravel envelope.

Withdrawals near the outcrop areas of aquifers are rather quickly balanced by increases in recharge and/or reductions in natural discharge. Withdrawals at significant distances downdip do not appreciably affect conditions in the outcrop area, and thus must be partly or largely supplied from water in storage in the aquifers and confining beds.

Problems

The reduction of storage in an aquifer in the vicinity of a pumping well is reflected in a decline in ground-water levels and is necessary in order to establish a hydraulic gradient toward the well. If withdrawals are continued for long periods in areas underlain by thick sequences of unconsolidated deposits, such as the Atlantic and Gulf Coastal Plain, the lowered ground-water levels in the aquifer may result in drainage of water from layers of silt and clay. The depletion of storage in fine-grained beds results in subsidence of the land surface.

The depletion of storage in confining beds is permanent, and subsidence of the land surface that results from such depletion is also permanent. However, depletion of storage in aquifers may not be fully permanent, depending on the availability of recharge. In arid and semiarid regions, recharge rates are extremely small, and depletion of aquifer storage is, for practical purposes, permanent.

Depletion of storage in the aquifers underlying large areas of the Atlantic and Gulf Coastal Plain is reflected in long-term declines in ground-water levels. The declines suggest that withdrawals in these areas are exceeding the long-term yield of the aquifers. This is a water-management problem that will become more important as rates of withdrawal and the lowering of water levels increase.

Another problem that affects ground-water development in the region concerns the presence of saline water in the deeper parts of most aquifers. The occurrence of saline water is controlled by the circulation of freshwater that becomes increasingly slow down the dip of the aquifers. Thus, in some of the deeper aquifers, the interface between freshwater and saltwater is inshore; in parts of the region, including parts of Long Island, New Jersey, and Mississippi, the interface in the most intensively developed aquifers is a significant distance offshore. Pumping near the interfaces has resulted locally in problems of saltwater encroachment.

SOUTHEAST COASTAL PLAIN

The Southeast Coastal Plain is an area of about 212,000 square kilometers in Alabama, Florida, Georgia, and South Carolina. It is a flat, low-lying area in which altitudes range from sea level at the coast to about 100 meters down the center of the Florida peninsula, and as much as 200 meters on hills in Georgia near the interior boundary of the region. Much of the area, including the Everglades in southern Florida, is a nearly flat plain less than 10 meters above sea level.

Geology

The land surface of the Southeast Coastal Plain is underlain by unconsolidated deposits of Pleistocene age consisting of sand, gravel, clay, and shell beds, and, in southeastern Florida, by semiconsolidated limestone. From the coast up to altitudes of nearly 100 meters, the surficial deposits are associated with marine terraces formed when the Coastal Plain was inundated at different times by the sea. In most of the region the surficial deposits rest on formation, primarily of middle to late Miocene age, composed of interbedded clay, sand, and limestone. The most extensive Miocene deposit is the Hawthorn Formation. The formations of middle to late Miocene age and, where those formations are absent, the surficial deposits overlie semiconsolidated limestones and dolomites that are as much as 1,500 meters thick. These carbonate rocks range in age from early Miocene to Paleocene and are referred to as Tertiary limestones.

Ground-Water Resources

The Tertiary limestone that underlies the Southeast Coastal Plain constitutes one of the most productive aquifers in the United States, and is the feature that justifies treatment of the region separately from the remainder of the Atlantic and Gulf Coastal Plain. The aquifer, which is known as the Floridan aquifer, underlies all of Florida and southeast Georgia and small areas in Alabama and South Carolina. The Floridan aquifer consists of layers several meters thick, composed largely of loose aggregations of shells of foraminifers and fragments of echinoids and other marine organisms interbedded with much thinner layers of cemented and cherty limestone. The Floridan, one of the most productive aquifers in the world, is the principal source of ground-water supplies in the Southeast Coastal Plain region.

Water supplies are obtained from the Floridan aquifer by installing casing through the overlying formations and drilling an open hole in the limestones and dolomites comprising the aquifer. The marked difference in ground-water conditions between the Southeast Coastal Plain and the Atlantic and Gulf Coastal Plain regions is apparent in the response of ground-water levels to withdrawals. In the Atlantic and Gulf region most large withdrawals are accompanied by a pronounced continuing decline in ground-water levels.

Problems

In southern Florida, south of Lake Okeechobee, and in a belt about 30 kilometers wide northward along the east coast of Florida to the vicinity of St. Augustine, the water in the Floridan aquifer contains more than 100 mg/l of chloride. In this area, most water supplies are obtained from surficial aquifers, the most notable of which underlies the southeastern part of Florida and in the Miami area consists of 30 to 100 meters of cavernous limestone and sand referred to as the Biscayne aquifer. The Biscayne is an unconfined aquifer recharged by local precipitation and by infiltration of water from canals that drain water from impoundments (conservation areas) developed in the Everglades. It is the principal source of water for municipal, industrial, and irrigation uses, and can yield as much as 5 m (3 $\min^{-1}$) (1,300 gal $\min^{-1}$) to small-diameter wells less than 25 meters deep, finished with open holes only 1 to 2 meters in length.

The surficial aquifers in the remainder of the region are composed primarily of sand, except in the coastal zones of Florida where the sand is interbedded with shells and thin limestones. These surficial aquifers serve as sources of small ground-water supplies throughout the region and are the primary sources of ground water where the water in the Floridan aquifer contains more than about 250 mg/l of chloride.

HAWAIIAN ISLANDS

The Hawaiian Islands region encompasses the state of Hawaii and consists of eight major islands occupying an area of 16,706 square kilometers in the Pacific Ocean, 3,700 kilometers southeast of California. The islands are the tops of volcanoes that rise from the ocean floor and stand at altitudes ranging from a few meters to more than 4,000 meters above sea level. Each island was formed by lava that issued from one or more eruption centers. The islands have a hilly to mountainous appearance, resulting from erosion that has carved valleys into the volcanoes and built narrow plains along parts of the coastal areas.

Geology

Each of the Hawaiian Islands is underlain by hundreds of distinct lava flows, composed mostly of basalt. The lavas issued in repeated outpourings from narrow zones of fissures, first below sea level, then above it. The lavas that extruded below the sea are impermeable. Those formed above sea level tend to be highly permeable, with interconnected openings that formed as the lava cooled, cavities and openings that were not filled by the overlying flow, and lava tubes (tunnels). The central parts of the thicker flows tend to be more massive and less permeable; the most common water-bearing openings are joints and faults that formed after the lava solidified. The lava flows in valleys and in parts of the coastal plains are covered by a thin layer of alluvium consisting of coral (limestone) fragments, sand-size fragments of basalt, and clay.

The fissures through which the lava erupted tend to cluster near eruption centers. Flows from the fissures moved down depressions on the adjacent slopes to form layers of lava that dip at angles of 4 to 10 degrees toward the margins of the volcanoes. The result, prior to modification by erosion, is a broad, roughly circular, gently convex mountain similar in shape to a warrior's shield. Thus, volcanoes of the Hawaiian type are referred to as shield volcanoes. When eruption along a fissure ceases, the lava remaining in the fissure solidifies to form a dike.

All of the islands have sunk, to some extent, as a result of a downward flexing of the earth's crust caused by the weight of the volcanoes. This has resulted in flows that formed above sea level being depressed below sea level. The upper parts of these flows contain freshwater that serves as an important source of water.

In mineral composition and nature of the water-bearing openings, the lavas that form the Hawaiian Islands are very similar to those in the Columbia Plateau region. Thus, from these two standpoints, the regions could be combined into one. There is, however, one important difference that justifies their treatment as separate regions. This difference relates to the presence of seawater around and beneath the islands, which significantly affects the occurrence and development of water supplies.

It is useful to divide the ground-water system of the Hawaiian Islands into three parts. The first part consists of the higher areas of the islands in the vicinity of the eruption centers. The rocks in these areas are formed into a complex series of vertical compartments surrounded by dikes developed along eruption fissures. The ground water in these compartments is referred to as dike-impounded water. The second and most important part of the system consists of the lava flows that flank the eruption centers and contain fresh ground water floating on saline ground water. These flank flows are partially isolated hydraulically from the vertical compartments developed by the dikes that surround the eruption centers. The fresh ground water in these flows is referred to as basal ground water. In parts of the coastal areas the basal water is confined by the overlying alluvium. The third part of the system consists of freshwater perched, primarily in lava flows, on soils, ash, or thick impermeable lava flows above basal ground water.

Some discharge of dike-impounded ground water occurs through fractures in the dikes into the flanking lava flows. This movement must be small, however, because water stands in the compartments at levels hundreds of meters above sea level, and the principal discharge occurs as springs on the sides and at the heads of valleys where erosion has removed parts of the dikes. Both the basal ground water and the perched ground water in the lava flows surrounding the dike-bounded compartments are recharged by precipitation and by streams leaving the dike-bounded area. Discharge is to streams and to springs and seeps along the coast.

The basal water is the principal source of ground water on the islands. Because the freshwater is lighter (less dense) than seawater, it floats as a lens-shaped body on the underlying seawater. The thickness of the freshwater zone below sea level essentially depends on the height of the freshwater head above sea level. Near the coast the zone is thin, but several kilometers inland from the coast on the larger islands it reaches thicknesses of at least a few hundred meters. In parts of the coastal zone, and especially on the leeward side of the islands, the basal ground water is brackish.

Approximately 50 percent of the water used in Hawaii is ground water. It is obtained through horizontal tunnels and through both vertical and inclined wells. Tunnels are used to obtain supplies of basal water near the coast where the freshwater zone is thin. Tunnels are used to tap dike-impounded water. These tunnels encounter large flows of water when the principal impounding dike is penetrated and it is necessary to drain most of the water in the saturated zone above the tunnel before construction can be completed. Thereafter, the yield of the tunnel reflects the rate of recharge to the compartment tapped by the tunnel. To avoid a large initial waste of water and to preserve as much storage as possible, the Honolulu Board of Water Supply has begun to construct inclined wells to obtain basal water and perched ground water in inland areas where the thickness of the freshwater zone permits the use of such wells.

ALASKA

The Alaska region encompasses the state of Alaska, which occupies an area of 1,519,000 square kilometers at the northwest corner of North America. Alaska can be divided into four divisions. From south to north, they are the Pacific Mountain System, the Intermontane Plateaus, the Rocky Mountain System, and the Arctic Coastal Plain. The Pacific Mountain System is the Alaskan equivalent of the Coast Range, Puget Sound Lowland, and Cascade provinces of the Washington-Oregon area. The Intermontane Plateaus is a lowland area of plains, plateaus, and low mountains comparable to the area between the Cascades-Sierra Nevada and the Rocky Mountains. The Rocky Mountain System is a continuation of the Rocky Mountains of the United States and Canada, and the Arctic Coastal Plain is the equivalent of the Great Plains of the United States and Canada. The coastal areas and lowlands range in altitude from sea level to about 300 meters, and the higher mountains reach altitudes of 1,500 to 3,000 meters. Mt. McKinley in the Pacific Mountain System is the highest peak in North America, with an altitude of about 6,200 meters.

Geology

As would be expected of any area its size, Alaska is underlain by a diverse assemblage of rocks. The principal mountain ranges have cores of igneous and metamorphic rocks ranging in age from Precambrian to Mesozoic. These are overlain and flanked by younger sedimentary and volcanic rocks. The sedimentary rocks include carbonates, sandstones, and shales. In much of the region the bedrock is overlain by unconsolidated deposits of gravel, sand silt, clay, and glacial till.

Ground-Water Resources

Climate has a dominant effect on hydrologic conditions in Alaska. Mean annual air temperatures range from $-12°C$ in the Rocky Mountain System and the Arctic Coastal Plain to about $5°C$ in the coastal zone adjacent to the Gulf of Alaska. The present climate and the colder climates that existed intermittently in the past have resulted in the formation of permafrost, or perenially frozen ground. Permafrost is present throughout the state except in a narrow strip along the southern and southeastern coasts. In the northern part of the Seward Peninsula, in the western and northern parts of the Rocky Mountain System, and in the Arctic Coastal Plain, the permafrost extends to depths as great as 600 meters, and is continuous except beneath deep lakes and in the alluvium beneath the deeper parts of the channels of streams. South of this area and north of the coastal strip, the permafrost is discontinuous and depends on exposure, slope, vegetation, and other factors. The permafrost is highly variable in thickness in this zone but is generally less than 100 meters thick.

Much of the water in Alaska is frozen for at least part of each year: on the surface as ice in streams and lakes, or as snow or glacier ice, and below the surface as winter frost and permafrost. Approximately half of Alaska, including the mountain ranges and adjacent parts of the lowlands, was covered by glaciers during the Pleistocene. About 73,000 square kilometers, or one-twentieth of the region, is still occupied by glaciers, most of which are in the mountain ranges that border the Gulf of Alaska. Precipitation ranges from about 130 mm yr^{-1} in the Rocky Mountain System and the Arctic Coastal Plain to about 7,600 mm yr^{-1} along the southeast coast. Precipitation falls as snow for 6 to 9 months of the year and even year-round in the high mountain regions. The snow remains on the surface until thawing conditions begin, in May in southern and central Alaska, and in June in the arctic zone. During the period of subfreezing temperatures, there is no overland runoff, and many streams and shallow lakes not receiving substantial ground-water discharge are frozen solid.

From the standpoint of ground-water availability and well yields, Alaska is divided into three zones. In the zone of continuous permafrost, ground water occurs beneath the permafrost and in small, isolated, thawed zones that penetrate the permafrost beneath large lakes and deep holes in the channels of streams. In the zone of discontinuous permafrost, ground water occurs below the permafrost and in sand and gravel deposits that underlie the channels and floodplains of major streams. In the zone of discontinuous permafrost, water contained in silt, clay, glacial till, and other fine-grained deposits usually is frozen. Thus, in this zone the occurrence of ground water is largely controlled by hydraulic conductivity. In the zone not affected by permafrost, which includes the Aleutian Islands, the western part of the Alaskan Peninsula, and the southern and southeastern coastal areas, ground water occurs in the bedrock and in the continuous layer of unconsolidated deposits that mantle the bedrock.

Little is known about the occurrence and availability of ground water in the bedrock. Permafrost extends into the bedrock in both the zones of continuous and discontinuous permafrost, but springs that issue from carbonate rocks in the Rocky Mountain System indicate the presence of productive water-bearing openings. Small supplies of ground water also have been developed from sandstones, from volcanic rocks, and from faults and fractures in the igneous and metamorphic rocks.

Recharge of the aquifers in the Alaska region occurs when the ground is thawed in the areas not underlain by permafrost. This period only lasts from June through September. Because the ground, even in nonpermafrost areas, is still frozen when most snowmelt runoff occurs, little recharge occurs in interstream areas by infiltration of water across the unsaturated zone. Instead, most recharge occurs through the channels of streams where they flow across the alluvial fans that fringe the mountainous areas and in alluvial deposits for some distance downstream. Because of the large hydraulic conductivity of the sand and gravel in these areas, the rate of infiltration is large.

Discharge from aquifers occurs in the downstream reaches of streams and through seeps and springs along the coast. The winter flow of most Alaskan streams is sustained by ground-water discharge. In the interior and northern regions, this discharge is evidenced by the buildup of ice (referred to locally as "icings") in the channels of streams and on the adjacent floodplains.

Problems

In the Arctic Coastal Plain and other areas underlain by very deep permafrost, ground water under the permafrost tends to contain a large concentration of dissolved substances. Objectionable concentrations of iron also are present in shallow aquifers in most parts of the region.

See table 11-1 for a breakdown of drilling costs by ground-water region.

Table 11-1. Cost of drilling in ten ground-water regions in dollars per foot.[a]

Region	Borehole Diameters							Aquifers	Maximum Yield (gpm)	Well Depth (ft)
	5	8	10	12	14	16				
Western Mountain Ranges	10.20	14.62	18.00	21.58	27.28	34.98		Alluvial	40,000	30–100
								Basalt	110	150–2,000
								Sandstone	20–1,500	100–2,000
								Limestone	Unknown	
Alluvial Basins	11.10	11.75	15.26	18.49		21.44		Alluvial	200–3,000	100–2,000
Columbia Lava Plateau	9.84	14.48	17.49	26.18		33.72		Basaltic Lavas	5–1,400	90–1,100
								Alluvial	2,500–3,500	30–150
Colorado Plateau and Wyoming Basin	8.92	11.00	14.74	22.23		27.98		Alluvial	500–1,000	60–300
								Interbedded Shale and Sandstone	25–500	120–700
								Limestone	Unknown	
High Plains[b]	4.50	5.10	6.30	7.50	9.74	12.00		Alluvial	300–30,000	50–700
								Interbedded Sandstone, limestone, shale	Unknown	

(continued)

Table 11-1. *(continued)*

Region	Borehole Diameters						Aquifers	Maximum Yield (gpm)	Well Depth (ft)
	5	8	10	12	14	16			
Nonglaciated Central Region	5.62	9.50	6.30	22.82			Alluvial Sandstone Limestone	55–2,000 1,000 Variable	30–200 120–1,500 40–600
Glaciated Central Region	8.52	11.50	19.39	27.41			Alluvial Sandstone	500–1,500 500–1,000	40–1,100 200–2,000
Piedmont and Blue Ridge Region	5.46	10.67	14.99	24.98			Metamorphic Limestone	20–750 4–10	300–500 200–500
Atlantic and Gulf Coastal Plain	6.24	9.50	14.09	22.27			Semiconsolidated Consolidated Sedimentary	150–1,200 10–6,500	10–800 100–500

a. Prices based on a recent well cost survey.
b. Cost of drilling using predominantly PVC casing.

Bibliography

Campbell, M. D., and J. H. Lehr. 1973. *Water Well Technology.* New York: McGraw-Hill.

Driscoll, F. 1986. *Groundwater and Wells.* 2nd ed. St. Paul, MN: Johnson Division, Signal Environmental Systems.

Driscoll, F. G., D. T. Hanson, and L. J. Page. 1980. Well-efficiency project yields energy-saving data, Parts 1–3. *Johnson Driller's Journal,* Mar./Apr., May/June, Sept./Oct.

Eggington, H. F., ed. 1985. *Australian Drillers Guide.* 2nd ed. NSW, Australia: Australian Drilling Industry Training Committee Limited.

The Engineers' Manual for Water Well Design. 1985. Los Angeles: Roscoe Moss Company.

Fetter, C. W., Jr. 1980. *Applied Hydrogeology.* Columbus, OH: Charles E. Merrill.

Gass, T. E., T. W. Bennett, J. Miller, and R. Miller. 1980. *Manual of Water Well Maintenance and Rehabilitation Technology.* Dublin, OH: National Water Well Association.

Gilluly, J., A. C. Waters, and A. O. Woodford. 1959. *Principles of Geology.* San Francisco: W. H. Freeman.

Goodrich, D. L. 1985. Step-drawdown and constant-rate pumping. *Water Well Journal* 39 (May): 39–42.

Hampton, E. R. 1964. Geologic factors that control the occurrence and availability of ground water in the Fort Rock Basin, Lake County, Oregon, *U.S. Geological Survey Professional Paper 383-B.*

Heath, R. C. 1984a. Basic ground water hydrology. *U.S. Geological Survey Water Supply Paper 2220.*

———. 1984b. Ground-water regions of the United States. *U.S. Geological Survey Water Supply Paper 2242.*

Heath, R. C., and F. W. Trainer. 1968. *Introduction to Ground Water Hydrology.* New York: Wiley.

International Hydrological Programme. 1984. *Ground water in hard rocks.* Project 8.6, prepared by the project panel, Ingemar Larsson, chairman. Paris: UNESCO.

Johnson Division, Universal Oil Products Company. 1960. Mud pressure aids cable tool drilling. *Johnson Driller's Journal* 32, no. 3 (May/June): 4–5.

Johnson Division, Universal Oil Products Company. 1966. *Ground Water and Wells.* 1st ed. St. Paul, MN: Universal Oil Products Company.

Johnson Division, Universal Oil Products Company. 1970. Self-sealing packer saves time on job. *Johnson Driller's Journal* 42, no. 6 (Nov./Dec.): 1–3.

Kozeny, J. 1933. Theorie und Berechnung der Brunnen, *Wasserkraft und Wasser Wirtschaft* 29:101.

Kruseman, G. P., and N. A. De Ridder. 1970. *Analysis and Evaluation of Pumping Test Data.* Wageningen, The Netherlands: International Institute for Land Reclamation and Improvement.

Lattman, L. H. 1958. Technique of mapping geologic fracture traces and lineaments on aerial photographs. *Photogrammetric Engineering* 84:568–76.

Lattman, L. H., and R. R. Parizek. 1964. Relationship between fracture traces and the occurrence of ground water in carbonate rocks. *Journal of Hydrology* 2:73–91.

Leet, L. D., S. Judson, and M. E. Kauffman. 1982. *Physical Geology.* Englewood Cliffs, NJ: Prentice-Hall.

Meinzer, O. E. 1923. The occurrence of ground water in the United States, *U.S. Geological Water Supply Paper 489.*

Meiser and Earl Hydrogeologists. 1982. *Use of Fracture Traces in Water Well Location: A Handbook.* Washington, D.C.: U.S. Office of Water Research and Technology.

National Water Well Association. 1979. *Water Well Drillers Beginning Training Manual.* Worthington, OH.

National Water Well Association of Australia. 1980. *Drillers Training and Reference Manual.* Macquarie Centre, New South Wales.

Newcomb, R. C. 1961. Storage of ground water behind subsurface dams in the Columbia River basalt, *U.S. Geological Survey Professional Paper 383-A.*

Press, F., and R. Siever. 1978. *Earth.* San Francisco: W. H. Freeman.

Safety Committee of the National Water Well Association. 1980. *Manual of Recommended Safe Operating Procedures and Guidelines for Water Well Contractors and Pump Installers.* Comp. H. W. Heiss, Jr. Dublin, OH: Water Well Journal Publishing Company.

Stallman, R. W. 1971. Aquifer test design, observation, and data analysis. *Techniques of Water Resources Investigations of the United States Geological Survey.* Washington, D.C.: GPO

Strahler, A. H. 1971. *The Earth Sciences.* New York: Harper & Row.

Thomas, H. E. 1952. *Ground Water Regions of the United States.* Report prepared for the House Committee on Interior and Insular Affairs. 83d Cong.

Todd, D. K. 1980. *Groundwater Hydrology.* New York: John Wiley.

UNICEF. 1985. Guidelines for drillers, engineers, geologists, and drilling trainees.

U.S. Environmental Protection Agency. 1975. *Manual of Water Well Construction Practices.* Washington, D.C.: Office of Water Supply.

Walton, W. C. 1962. Selected analytical methods for well and aquifer evaluation. *Illinois State Water Survey Bulletin 49*, Urbana, p. 81.

Index

Acids, 174
Acoustic water level device, 144
Aerial photographs, 13, 183–86
Air lifting, 158–59
Airline method, 143
Air-percussion rotary drilling, 52–55
Air rotary drilling, 45–46
Air surging, 159
Alaska, 217–19
Alignment, 118
Alluvial Basins, 195–96
Annular space, 125–26
Aquiclude, 4
Aquifer boundaries, 134
Aquifers, 4
  consolidated nature of, 7–8
  exploration for, 13–16
  properties of, 8–13
  rock types, 6–8
  unconfined vs. confined, 5–6
Aquifer test, 133, 145–46
Artesian well, 6
Atlantic and Gulf Coastal Plain, 211–13
Auger-bucket drilling, 60–61
Axial flow (water), 77

Bacterial incrustation, 174
Bail-down method, 117
Bailer, 31
Beart, Robert, 35
Bedrock, 187
Bentonite, 123
Biocides, 174
Biological incrustation, 164–65
Blasting, 192
Blocking, 106
Borehole, 14, 15, 49–50, 110
Borehole wall, 149, 151
Bridge slot, 88–89

Brule Formation, 201
Brushing, 191

Cable tool drilling
  advantages of, 33–34
  basics of, 26–27
  disadvantages of, 34
  rigs, 27–34
Caliche, 202
Caliper logs, 16
Capillary zone, 3
Capping (well), 132
Carbonate deposition, 162–63
Carbon dioxide, 163
Casing, 31–32, 73, 187
  method, 127
  selection of, 79–83
Cavings, 70
Cement slurry, 123
Centrifugal pump, 49
Charting, 94
Chemical corrosion, 86, 167
Chemical incrustation, 162–64, 174
Chemicals, 174–77
Chlorine solution, 130
Church, Melvin, 32
Circular orifice weir, 138
Collapse forces, 82
Colorado Plateau and Wyoming Basin, 198–200
Columbia Lava Plateau, 196–98
Column loading factor, 81–82
Cone of depression, 22
Confined aquifer, 6, 102–4
Confining bed, 4, 134
Consolidated crystalline aquifers, 8
Consolidated rock-well design, 178–79
  drilling methods, 187–90
  exploration techniques, 180–87

225

fracture characteristics, 179–80
  pump testing, 190–91
Consolidated sedimentary aquifers, 8
Constant rate test, 135
Converging flow, 21–22
Core samples, 66
Corrosion, 167–68
Costs (drilling), 178
Cutting samples, 66

Darcy, Henri, 19
Darcy's law, 19–20
Dart-valve bailer, 69
Data analysis, 146–47
Development, 127–29. *See also* Water-well development
Dewatering well, 75
Diameter (well), 76–79
Disinfection, 129–30
Distance drawdown graph, 147
Dolomite, 180
Drag bit, 49
Drake, E. L., 26
Drawdown, 21–22
Drawdown seal, 166
Drill bit, 29–30, 36, 49, 57
Drilled pipe, 88
Drilling fluid, 38–45
Drilling jar, 28–29
Drilling mud, 40, 111–12
Drilling technology, 25
  cable-tool drilling, 26–34
  rotary drilling, 34–45
  sampling methodology, 63–71
  special application systems, 56–63
Drill pipe, 37, 55
  double-walled, 51
Drill stem, 29, 37
Drive-core sampling, 69
Driven wells, 61–63
Drive shoe, 32
Dual-tube rotary drilling, 51–52

Eductor pipe, 159
Effective size, 90, 93–94
Efficiency, 84–85
Electrical resistivity surveys, 15, 187
Electrical shock hazards, 107

Electrical tape method, 143
Electrochemical corrosion, 167
Energy intensity, 151–52
Eolian, 203
Equilibrium, 23
Exploration techniques, 180–87
Extrusive igneous rock, 6–7

Fauvelle, 34–35
Field data, 181–82
Filter cake, 41–42
Filter design, 90–91
Filter pack, 150
Filter zone, 149
Fines, 150
Finishing (site), 130–32
Finishing (well), 127–30
Floridan aquifer, 214
Foam drilling, 47
Formation samples, 67–68
Fractures, 179–80
Fracture trace analysis, 183–85
Fracture zones, 189
Freshwater, 212–13

Gamma ray logs, 16
Geologic maps, 13, 182
Geologic reports, 181–82
Geophysical exploration, 14, 187
Geophysical logs, 15–16
Glacial outwash, 209
Glaciated Central Region, 205–6
Graded grain pack, 119
Gravel pack, 73, 90
  installation of, 119–21
Gravel-packed wells, 95–100
Gravity placement, 125
Ground water, 1
Ground-water flow, 17–21
Ground-water regions
  Alaska, 217–19
  Alluvial Basins, 195–96
  Atlantic and Gulf Coastal Plain, 211–13
  classification of, 193
  Colorado Plateau and Wyoming Basin, 198–200
  Columbia Lava Plateau, 196–98
  Glaciated Central Region, 205–6

Hawaiian Islands, 214–17
High Plains, 200–202
Nonglaciated Central Region,
    202–5
Northeast and Superior Uplands,
    209–11
Piedmont Blue Ridge Region,
    207–9
Southeast Coastal Plain, 213–14
Western Mountain Ranges,
    193–95
Grout, 73, 121–24
    casing method, 127
    mixing, 124–25
    piping, 126–27
    placement of, 125–26

Hawaiian Islands, 214–17
Head loss, 84, 85
Heath, R. C., 193
Heaving, 68
Heterogeneous aquifers, 94–95
High Plains, 200–202
High velocity jetting, 159–60
Hollow-rod drilling, 59–60
Homemade openings, 87–88
Homogeneous aquifers, 93–94
Hydraulic conductivity, 10–13, 17
Hydraulic gradient, 18–20
Hydraulic head, 18
Hydrofracturing, 192
Hydrologic cycle, 1
Hydrologist, 64

Igneous rock, 6
Incrustation, 86, 162–66
    problem solving, 174
Intake design, 84–85
    filter design, 90–91
    gravel-packed wells, 95–100
    installation methods, 114–18
    intake length, 100–104
    naturally-developed well design,
        91–95
    physical properties of, 85–87
    types of intakes, 87–90
Intrusive igneous rock, 6
Iron bacteria, 165
Iron oxides, 163–64

Jet drilling, 56–59
Johnson prototype, 144

Kelly, Henry, 26
Kelly (rotary drilling), 38
Kozeny's curves, 102–3

Lag factor, 70
Laminar flow, 17
Limestone, 12–13, 180
Lineaments, 183
Line scanning, 14
Loess, 205
Log, 15
Louvered slots, 89

Machine-cut slots, 88
Magma, 6
Marsh funnel, 44–45
Meinzer, O. E., 193
Metamorphic rocks, 7
Mud, 39
    balance, 44
    density of, 43–44
Multiple-step drawdown test, 135

Naturally developed wells, 91–95
Negative boundaries, 23
Nonglaciated Central Region, 202–5
Northeast and Superior Uplands,
    209–11

Observation wells, 145–46
Ogallala Formation, 201
Orange peel bucket, 51
Overpumping, 154–55
Oxygen, 166

Packing, 10
Perched aquifer, 4
Permafrost, 218
Permeability, 150
Permits, 105–6
Petroleum industry, 64
pH levels, 42–43
Phosphates, 174
Physical incrustation, 164–65, 174
Piedmont Blue Ridge Region, 207–9
Pilot hole, 112

Plumbness, 118
Porosity, 8–10, 17, 150
Positive boundaries, 23
Potentiometric surface maps, 6
Power lines, 107
Pull-back method, 114–16
Pump house, 132
Pumping system, 75–76
Pumping test, 190–91
Pumping water level, 22
Punched intake openings, 88–89
PVC casing, 113

Radius of influence, 22
Rawhiding, 155–56
Recharge, 3, 22–23
Recharge boundary, 134
Recovery test, 136
Registration, 105–6
Regolith, 207
Remote sensing, 13–14
Resistivity logs, 16
Return flow method, 69
Reverse-circulation drilling, 47–51
Rock roller bits, 49
Rocks, 64
Rock types, 6–8
Rope socket, 28
Rotary drilling
   air-percussion method, 52–55
   air rotary, 45–46
   components of, 36–38
   drilling fluids, 38–45
   dual-tube method, 51–52
   foam drilling, 47
   history of, 34–36
   reverse-circulation drilling, 47–51
Rotary-table drive, 46

Safety, 106–9
Saline water, 212–13
Samples, 65
Sample splitters, 66
Sampling methodology, 63–67
   formation samples, 67–68
   methods of taking, 69–71
Sand bridging, 154–55
Sand-pump bailer, 69
Sandstone, 180

Saprolite, 207
Satellite photographs, 13, 186
Saturated zone, 2
Saw-cut slots, 87–88
Screen, 73
Seal, 73
Seasonally operated wells, 173–74
Sedimentary rocks, 7
Seismic surveys, 14–15, 187
Self-sealing packer, 116
Semiconsolidated sedimentary
   aquifer, 7
Settling pits, 66
Site preparation, 106–9
Slot size, 92
Slug test, 135
Soil water belt, 2
Sorting, 10
Southeast Coastal Plain, 213–14
Specific capacity, 76–77
Specific retention, 9
Specific yield, 8–9
Splitting of samples, 65
Spontaneous potential logs, 16
Springs, 4
Spudding beam, 30
Stabilizing pack, 99–100
Static water level, 21, 156
Sulfate-reducing bacteria, 165
Surge block, 156–58
Surging, 156–58

Tectonic forces, 179, 182
Telescoping well screens, 115
Tensile forces, 80–81
Test drilling, 74
Testing
   data analysis phase, 146–47
   kinds of tests, 134–36
   observation phase, 138–46
   planning phase, 136–38
   preliminary studies, 133–34
Threads, 27–28, 37
Till, 209
Time drawdown graph, 147
Tool string, 27
Top-head drive rig, 46
Topographic maps, 182
Torch-cut slots, 87

Transmissivity, 11, 21
Transpiration, 2
Tremie pipe method, 120
Troubleshooting, 168-71
Tungsten carbide buttons, 54
Turbulent flow, 17
Turning clamp, 57

Unconfined aquifer, 5, 102
Unconsolidated alluvial aquifer, 7
Uniform grain pack, 119
Uniformity coefficient, 90-92, 93-94
Unsaturated zone, 2

Vadose zone, 2
Valley, 195
Velocity (water), 78
Viscosity (drilling fluid), 44-45
Volumetric method, 138

Wash-down method, 117-18
Water, 42-45
  axial flow of, 77
  loss of, 50
  velocity of, 78
  volumetric method, 138
  water level measurement, 139, 142-45
Water-bearing strata, 111-13
Water-cement ratio, 124
Water resource maps, 182
Water resource reports, 182
Water table, 2-3, 18
Water-well construction
  equipment coordination, 113-14
  finishing, 127-30

  gravel-pack installation, 119-21
  grouting and sealing, 121-27
  installation preparations, 110-11
  intake installation methods, 114-18
  permits and registration, 105-6
  plumbness and alignment, 118
  site finishing, 130-32
  site preparation and safety, 106-9
  water-bearing strata recognition, 111-13
Water-well design
  casing type selection, 79-83
  diameter, 76-79
  initial design information, 74-75
  water wells, 72-73
  yield determination, 75-76
Water-well development
  air lifting, 158-59
  air surging, 159
  high velocity jetting, 159-60
  importance of, 148-50
  overpumping, 154-55
  principles of, 151-53
  rawhiding, 155-56
  surging, 156-58
Water-well maintenance
  corrosion, 167-68
  incrustation, 162-66
  solutions to problems, 172-77
  troubleshooting, 168-71
  well failure, 161-62
Well failure, 161-62
Well record, 64
Western Mountain Ranges, 193-95
Wetted tape method, 142-43
Wire-wrapped screen, 89-90